Energy, Climate and the Environment

Series Editor: **David Elliott**, Emeritus Professor of Technology Policy, Open University, UK

Titles include:

Espen Moe
RENEWABLE ENERGY TRANSFORMATION OR FOSSIL FUEL BACKLASH
Vested Interests in the Political Economy

Manuela Achilles and Dana Elzey (*editors*)
ENVIRONMENTAL SUSTAINABILITY IN TRANSATLANTIC PERSPECTIVE
A Multidisciplinary Approach

Robert Ackrill and Adrian Kay (*editors*)
THE GROWTH OF BIOFUELS IN THE 21ST CENTURY

Philip Andrews-Speed
THE GOVERNANCE OF ENERGY IN CHINA
Implications for Future Sustainability

Gawdat Bahgat
ALTERNATIVE ENERGY IN THE MIDDLE EAST

Ian Bailey and Hugh Compston (*editors*)
FEELING THE HEAT
The Politics of Climate Policy in Rapidly Industrializing Countries

Mehmet Efe Biresselioglu
EUROPEAN ENERGY SECURITY
Turkey's Future Role and Impact

Jonas Dreger
THE EUROPEAN COMMISSION'S ENERGY AND CLIMATE POLICY
A Climate for Expertise?

Beth Edmondson and Stuart Levy
CLIMATE CHANGE AND ORDER
The End of Prosperity and Democracy

David Elliott (*editor*)
NUCLEAR OR NOT?
Does Nuclear Power Have a Place in a Sustainable Future?

Neil E. Harrison and John Mikler (*editors*)
CLIMATE INNOVATION
Liberal Capitalism and Climate Change

Antonio Marquina (*editor*)
GLOBAL WARMING AND CLIMATE CHANGE
Prospects and Policies in Asia and Europe

Espen Moe and Paul Midford (*editors*)
THE POLITICAL ECONOMY OF RENEWABLE ENERGY AND ENERGY SECURITY
Common Challenges and National Responses in Japan, China and
Northern Europe

Marlyne Sahakian
KEEPING COOL IN SOUTHEAST ASIA
Energy Consumption and Urban Air-Conditioning

Benjamin K. Sovacool
ENERGY & ETHICS
Justice and the Global Energy Challenge

Joseph Szarka, Richard Cowell, Geraint Ellis, Peter A. Strachan and
Charles Warren (*editors*)
LEARNING FROM WIND POWER
Governance, Societal and Policy Perspectives on Sustainable Energy

Thijs Van de Graaf
THE POLITICS AND INSTITUTIONS OF GLOBAL ENERGY GOVERNANCE

Xu Yi-chong (*editor*)
NUCLEAR ENERGY DEVELOPMENT IN ASIA
Problems and Prospects

Claire Dupont and Sebastian Oberthür (*editors*)
DECARBONIZATION IN THE EUROPEAN UNION
Internal Policies and External Strategies

Energy, Climate and the Environment
Series Standing Order ISBN 978–0–230–00800–7 (hb) 978–0–230–22150–5 (pb)
(*outside North America only*)

You can receive future titles in this series as they are published by placing a
standing order. Please contact your bookseller or, in case of difficulty, write to
us at the address below with your name and address, the title of the series and
the ISBNs quoted above.

Customer Services Department, Macmillan Distribution Ltd, Houndmills,
Basingstoke, Hampshire RG21 6XS, UK.

Decarbonization in the European Union

Internal Policies and External Strategies

Edited by

Claire Dupont
Senior Researcher, Institute for European Studies, Vrije Universiteit Brussel, Belgium

and

Sebastian Oberthür
Academic Director, Institute for European Studies, Vrije Universiteit Brussel, Belgium

First published 2015 by
PALGRAVE MACMILLAN

Palgrave Macmillan in the UK is an imprint of Macmillan Publishers Limited,
registered in England, company number 785998, of Houndmills, Basingstoke,
Hampshire RG21 6XS.

Palgrave Macmillan in the US is a division of St Martin's Press LLC,
175 Fifth Avenue, New York, NY 10010.

Palgrave Macmillan is the global academic imprint of the above companies
and has companies and representatives throughout the world.

Palgrave® and Macmillan® are registered trademarks in the United States,
the United Kingdom, Europe and other countries.

ISBN 978–1–137–40682–8

This book is printed on paper suitable for recycling and made from fully
managed and sustained forest sources. Logging, pulping and manufacturing
processes are expected to conform to the environmental regulations of the
country of origin.

A catalogue record for this book is available from the British Library.

A catalog record for this book is available from the Library of Congress.

In memoriam Marc Pallemaerts

This volume is dedicated to Marc Pallemaerts. It was in the middle of this project that the news of Marc's sudden and unexpected death on 2 May 2014 reached us and left us in shock. We cannot possibly do justice here to Marc's rich contribution to environmental law and policy in Belgium and internationally in scholarly, societal and political terms. For glimpses of his contribution, we refer the reader to a number of tributes that are already available, including http://www .klimaat.be/nl-be/news/20142/hommage-aan-marc-pallemaerts; http:// www.ieep.eu/about-us/in-memory-of-marc-pallemaerts/; http://www.om gevingsrecht.be/opinie/hommage-aan-marc-pallemaerts. Here, we wish to pay tribute to Marc's role as a respected colleague and cherished friend during the early years at the Institute for European Studies (IES), Vrije Universiteit Brussel (VUB), and afterwards.

Marc was among the very first staff of the IES, and he served as a Senior Research Fellow from 2002 to late 2005. As such, he helped set up and shape the Institute, including, in particular, its environmental arm that since has taken the form of the IES research cluster on Environment and Sustainable Development (http://www.ies.be/ research/EnvironmentandSustainableDevelopment). He was also instrumental in attracting the Institute's first Academic Director (whom he encouraged to apply) and in framing the IES research strategy, valid to the present day, focusing on Europe and the European Union (EU) in an international context.

Collaboration continued after Marc left the IES to become a Senior Fellow at the Institute for European Environmental Policy (IEEP). He was the driving force behind the establishment of the joint IES–IEEP Environmental Policy Forum, which has since grown into one of the main event outlets of the Institute, the IES Policy Forum (http://www.ies.be/ events/policy-forums). He also remained involved in the supervision of several IES PhD students. The editors of this book furthermore look back at the fruitful cooperation with Marc on the very successful publication of the edited volume *The New Climate Policies of the European Union: Internal Legislation and Climate Diplomacy* (2010), which itself emanated from cooperation on a lecture series on 'The European Union and the Fight against Global Climate Change' in autumn 2008.

During all these years, we cherished Marc's unwavering straightforwardness and his uncompromising dedication to core values, including the environment and human rights. By dedicating this book to him, we hope to contribute to keeping his memory and commitment alive. We are confident that he would have liked the analysis of this book, although it took more time to complete than he, as it turned out, had. Marc's contribution and influence are still felt and appreciated, including at the IES. We miss him as a colleague and friend. He was, and will remain, a strong source of inspiration and encouragement for many at the IES and beyond.

The Institute for European Studies, Brussels, February 2015

Bart De Schutter	Sebastian Oberthür	Claire Dupont
President	Academic Director	Senior Researcher

Contents

Tables and Figures

Tables

Figures

ix

Series Editor's Preface

Concerns about the potential environmental, social and economic impacts of climate change have led to a major international debate over what could and should be done to reduce emissions of greenhouse gases. There is still a scientific debate over the likely *scale* of climate change and the complex interactions between human activities and climate systems, but global average temperatures have risen and the cause is almost certainly the observed build-up of atmospheric greenhouse gases.

Whatever we now do, there will have to be a lot of social and economic adaptations to climate change – preparing for increased flooding and other climate-related problems. However, the more fundamental response is to try to reduce or avoid the human activities that are causing climate change. That means, primarily, trying to reduce or eliminate emission of greenhouse gases from the combustion of fossil fuels. Given that around 80 per cent of the energy used in the world at present comes from these sources, this will be a major technological, economic and political undertaking. It will involve reducing demand for energy (via lifestyle choice changes – and policies enabling such choices to be made), producing and using whatever energy we still need more efficiently (getting more from less) and supplying the reduced amount of energy from non-fossil sources (basically switching over to renewables and/or nuclear power).

Each of these options opens up a range of social, economic and environmental issues. Industrial society and modern consumer cultures have been based on the ever-expanding use of fossil fuels, so the changes required will inevitably be challenging. Perhaps equally inevitable are disagreements and conflicts over the merits and demerits of the various options and in relation to strategies and policies for pursuing them. These conflicts and associated debates sometimes concern technical issues, but there are usually also underlying political and ideological commitments and agendas which shape, or at least colour, the ostensibly technical debates. In particular, at times, technical assertions can be used to buttress specific policy frameworks in ways which subsequently prove to be flawed.

The aim of this series is to provide texts which lay out the technical, environmental and political issues relating to the various proposed policies for responding to climate change. The focus is not primarily on the science of climate change, or on the technological detail, although there will be accounts of the state of the art, to aid assessment of the viability of the various options. However, the main focus is the policy conflicts over which strategy to pursue. The series adopts a critical approach and attempts to identify flaws in emerging policies, propositions and assertions. In particular, it seeks to illuminate counter-intuitive assessments, conclusions and new perspectives. The aim is not simply to map the debates but to explore their structure, their underlying assumptions and their limitations. Texts are incisive and authoritative sources of critical analysis and commentary, indicating clearly the divergent views that have emerged and also identifying the shortcomings of these views. However, the books do not simply provide an overview, they also offer policy prescriptions.

That is certainly the case for the present volume, which looks at how the EU might manage its evolving internal and external energy relations. It reviews how decarbonization policies have emerged within the EU and explores the internal barriers to achieving full climate policy integration, with overviews of the issues in specific sectors. It also looks at how EU decarbonization will affect the geopolitics of energy and, in particular, EU relations with major partner countries and neighbours, including Norway and Russia. It takes a long-term policy perspective, towards 2050, arguing that this enables a strategic approach to be taken. The current flurry of concern over relations with Russia, focused in energy terms mainly on short-term fossil fuel access and pricing issues, adds urgency to this task – we need to look further ahead and develop a more coherent approach, linking internal and external policies. This book offers some good starting points.

Acknowledgements

This book results from an exploration of the theme of decarbonization during a lecture series entitled 'EU Energy Policy: On the Way to Decarbonization?' which was organized by the Institute for European Studies (IES) at the Vrije Universiteit Brussel (VUB) in collaboration with Climate Strategies, WWF and the Ecologic Institute in the autumn of 2012. The ten public lectures (which can be viewed online at http://www.ies.be/node/1364) touched on topics of internal EU policy development to combat climate change and how external relations could develop as a result of decarbonization. What was striking about this lecture series was that both the speakers and the audience raised many questions about how to achieve decarbonization, what it would mean for the European society and how it would affect international relations. It soon became clear that these questions deserved further attention. Following on from this lecture series, we were thus delighted to find a group of scholars interested in such a research agenda and in working with us in developing this book to provide a more in-depth exploration of these issues. This has proved an enriching exercise, which will certainly lead to further and more nuanced research and debate.

We are particularly grateful for the financial support received under the European Commission's Jean Monnet Lifelong Learning Programme for both the organization of the lecture series and the publication of this book. The research and editing work for this book was also supported by the VUB's strategic research programme 'Evaluating Democratic Governance in Europe (EDGE)' (http://www.edge-programme.eu/), pursued jointly by the VUB's Political Science Department and the IES.

We would also like to thank the staff of the IES for their hard work in organizing the lecture series, Arianna Khatchadourian and Jolyon Larson for editorial assistance and the contributing authors to the volume for their interest in pursuing this research agenda and for their timely responses to our requests and comments.

Contributors

Max Åhman is Associate Professor of Environmental and Energy Systems Studies at Lund University, Sweden.

Elin Lerum Boasson is a senior research fellow at the Center for International Climate and Environmental Research (CICERO), Oslo, Norway.

Tom Casier is Jean Monnet Chair, Academic Director and Senior Lecturer in International Relations at the Brussels School of International Studies, University of Kent, Brussels, Belgium.

Claire Dupont is a post-doctoral researcher under the 'Evaluating Democratic Governance in Europe' (EDGE) research programme at the Institute for European Studies and the Political Science Department, Vrije Universiteit Brussel, Belgium.

Torbjørg Jevnaker is a research fellow at the Fridtjof Nansen Institute, Lysaker, Norway.

Olga Khrushcheva is a lecturer at the Manchester Metropolitan University, UK.

Stefan Lechtenböhmer is Director of the research group 'Future Energy and Mobility Structures' at the Wuppertal Institute for Climate, Environment and Energy, Germany.

Leiv Lunde is Director of Strategy at the Norwegian Ministry of Foreign Affairs.

Cathy Macharis is a professor at the Vrije Universiteit Brussel, Belgium, leading the Mobility, Logistics and Automotive Technology Research Centre (MOBI), and visiting professor at the University of Gothenborg, Sweden.

Tomas Maltby is Lecturer in International Politics at King's College London, UK.

Lars J. Nilsson is Professor of Environmental and Energy Systems Studies, Lund University, Sweden.

Sebastian Oberthür is the Academic Director of the Institute for European Studies, Vrije Universiteit Brussel, Belgium, with a research focus on European and international climate and environmental governance.

Radostina Primova is an EU Affairs Consultant at Hinicio and an associate researcher at the Institute for European Studies, Vrije Universiteit Brussel, Belgium.

Sascha Samadi is a research fellow in the research group 'Future Energy and Mobility Structures' at the Wuppertal Institute for Climate, Environment and Energy, Germany.

Thomas Sattich is an associate researcher at the Institute for European Studies, Vrije Universiteit Brussel, Belgium.

Jon Birger Skjærseth is a research professor at the Fridtjof Nansen Institute, Lysaker, Norway.

Tom van Lier is a research associate at the Mobility, Logistics and Automotive Technology Research Centre (MOBI), Vrije Universiteit Brussel, Belgium.

Acronyms and Abbreviations

ACER	EU Agency for the Cooperation of Energy Regulators
BAU	Business As Usual
bcm	Billion cubic metres
°C	Degrees Celsius
CCS	Carbon Capture and Storage
CO	Carbon monoxide
CO_2	Carbon dioxide
CO_2eq	Carbon dioxide equivalent
DG	Directorate General
ECF	European Climate Foundation
EEA	European Environment Agency *or* European Economic Area
EED	Energy Efficiency Directive
EII	Energy Intensive Industry
EFTA	European Free Trade Area
ENPI	European Neighbourhood Partnership Instrument
ENTSO-E	European Network of Transmission System Operators for Electricity
EPBD	Energy Performance of Buildings Directive
EREC	European Renewable Energy Council
ETS	Emissions Trading System
EU	European Union
FDI	Foreign Direct Investment
GDP	Gross Domestic Product
GHG	Greenhouse Gas
H_2	Hydrogen
ICT	Information and Communications Technology
IPCC	Intergovernmental Panel on Climate Change
LNG	Liquefied Natural Gas
MENA	Middle East and North Africa
MEP	Member of the European Parliament
Mt	Megatonnes
MWh	Megawatt hours
MWh/km^2	Megawatt hours per kilometre squared
NER300	New Entrants Reserve

NGO	Non-Governmental Organization
OPEC	Organization of the Petroleum Exporting Countries
OSCE	Organization for Security and Cooperation in Europe
PV	Photovoltaic
QMV	Qualified Majority Voting
R&D	Research and Development
RE	Renewable Energy
RES	Renewable Energy Sources
SET-plan	Strategic Energy Technology plan
SME	Small and Medium-sized Enterprise
TAP	Trans-Adriatic Pipeline
TEN-E	Trans-European Network for Energy
TFEU	Treaty on the Functioning of the European Union
TWh	Terawatt Hours
UNFCCC	United Nations Framework Convention on Climate Change
VAT	Value-Added Tax
WTO	World Trade Organization
WWF	Worldwide Fund for Nature

1
Decarbonization in the EU: Setting the Scene

Claire Dupont and Sebastian Oberthür

Introduction

Climate change is a challenge that requires long-term, cross-sectoral and cross-border action. It is often described as one of the most complex problems facing humankind – one that affects the entire planet and all aspects of modern society (Haug et al., 2010). The European Union (EU) is not immune to the challenges of mitigating and adapting to climate change. Human-induced climate change is caused by the emission of potent and long-lived greenhouse gases (GHGs), which has increased since pre-industrial times (IPCC, 2013). Reducing GHG emissions quickly enough to a level that will prevent 'dangerous anthropogenic interference with the climate system' (UNFCCC, Article 2) means transforming our infrastructure, energy, transport, agriculture and industrial sectors away from fossil fuels.

The EU has long-held ambitions to demonstrate global leadership by example in responding to climate change, including through developing ambitious domestic policies (for example, Oberthür & Roche Kelly, 2008; Schreurs & Tiberghien, 2007). In October 2009, the European Council of heads of state and government agreed on an ambitious, long-term climate policy objective to demonstrate the EU's willingness to play a leading role in limiting global temperature increases to 2°Celsius (European Council, 2009). This agreed objective is to reduce GHG emissions in the EU by 80–95 per cent by 2050, compared to 1990 levels. Such an objective is in line with scientific estimates of the required action to avoid catastrophic climate change (IPCC, 2007; 2013).

Reducing GHG emissions to such a degree effectively means decarbonizing the EU. It requires eliminating emissions in a number of sectors of society and limiting emissions considerably elsewhere. As it especially requires a virtual phase-out of carbon dioxide emissions from the use of fossil fuels, this ambition is therefore often referred to as 'decarbonization'. Throughout this book, it is in the context of the objective to reduce GHG emissions in the EU by 80–95 per cent by 2050 that we use the term 'decarbonization'.

This volume explores what the 2050 decarbonization objective means in practice: first, for internal policies within the EU, and second, for external EU energy relations. Authors explore the EU's internal policies and external energy relations to see if they are in line with climate objectives to 2050. Are internal policies equipped to move towards and achieve decarbonization? Are external energy relations prepared for the transition that is unfolding? The challenges arising from decarbonization are only amplified by the contexts of economic and financial crises in Europe from 2008 onwards, political tension with Russia from 2013 onwards and changing geopolitics, given the rise of powers such as China, India, Brazil, among others. This book provides an assessment of the state of the art of EU internal policies and external relations and discusses how decarbonization should or could influence their development.

In this introductory chapter, we briefly discuss the historical development of EU climate policy in line with international climate developments. Next, we present the 2050 perspective, the nature of the 2050 objective and the many scenarios and roadmaps that describe options for achieving the decarbonization objective. We then introduce the key questions and accompanying conceptual framework guiding the contributions to the book. Finally, we provide an overview of the organization of the book and its chapters.

1. Development of EU climate policy

EU policy on climate change first developed in response to international developments. It was not until the 1980s that EU institutions began seriously considering climate change as an area for internal policy development – largely in response to international negotiations that eventually led to agreement on the United Nations Framework Convention on Climate Change (UNFCCC) in 1992. EU climate policy was often based on legal competences in the areas of the environment and the internal market. These competences allowed EU institutions to make progress on climate policy relatively quickly after the issue came onto the agenda (see, also for the following, Jordan et al., 2010; Oberthür & Pallemaerts, 2010; Wurzel & Connelly, 2011).

Throughout the 1990s, the EU moved forward in small steps on climate policy. In the international negotiations on what became the Kyoto Protocol to the UNFCCC in 1997, the EU proposed to reduce its GHG emissions by up to 15 per cent compared to 1990 levels by 2010. While this target was eventually lowered to a GHG emission reduction of 8 per cent under the Kyoto Protocol, this was still the most ambitious commitment by any party to the Protocol (Oberthür & Pallemaerts, 2010). Several EU climate-related policy measures were agreed in the 1990s and first half of the 2000s, including on GHG emissions monitoring (Council Decision 93/389/500/EEC and subsequent revisions in 1999 and 2004); energy efficiency (for example, Council Directive 93/79/EEC to improve energy efficiency, 'SAVE'; see Chapter 7); renewable energy (Council Decision 93/500/EC on the promotion of renewable energy sources, 'ALTENER' and Directive 2001/77/EC); emissions trading (Directive 2003/87/EC); and reducing emissions of fluorinated GHGs (Directive 2006/40/EC and Regulation EC 842/2006).

In the mid-2000s, the pace of internal EU climate policy development quickened, spurred on by several factors. First, climate policy became a driver of European integration more broadly. In the wake of the failed EU Constitutional Treaty, climate policy was reframed as an opportunity to reinforce the legitimacy of the EU and its institutions. Second, discussions on the security of energy supplies, given Europe's great energy import dependence (see also Chapter 8), provided an added impetus for more climate policies that would lead to enhanced domestic generation of renewable energy and increased energy efficiency. Third, the role of the EU in the international system and its strong support for multilateralism also provided added motivation to advance on climate policy (Roche Kelly et al., 2010, pp. 14–15).

Climate change was thus a matter of high politics by the time the EU came to agreeing internal policies to 2020 in 2007. Ambitions for 2020 were summarized in the 20-20-20 commitment: to reduce GHG emissions in the EU by 20 per cent by 2020 compared to 1990 levels; to increase the share of renewable energy in EU final energy consumption to 20 per cent by 2020; and to improve energy efficiency in the EU by 20 per cent compared to business-as-usual projections for 2020. Of these three goals, only the first two are binding. The energy efficiency target is an aspirational goal, although later policy measures were agreed to try to achieve the target (such as the Energy Efficiency Directive 2012/27/EU; see also Chapter 7). Adopted in 2009, the package of policy measures aiming to achieve the 2020 goals included a revised Emissions Trading Directive (2009/29/EC), a new Renewable Energy

Directive (2009/28/EC), a Directive providing a legal framework for Carbon Capture and Storage (CCS) technology (2009/31/EC) and a new Effort Sharing Decision (No. 406/2009/EC), allocating the amount of emissions each member state must reduce in sectors not covered by the Emissions Trading System (ETS). A regulation reducing the emissions of carbon dioxide from new passenger cars (No. 443/2009) was also negotiated and agreed alongside the climate and energy package, as was the inclusion of international aviation in the ETS (Directive 2008/101/EC). The 'climate and energy package' of 2008/09 contributed to proclamations that, perhaps, the EU had finally achieved a high level of integration of climate and energy policies (see, for example, Adelle et al., 2012). The package was eventually reflected in the Doha Amendment to the Kyoto Protocol, adopted in 2012, in which the EU agreed to reducing its GHG emissions by 20 per cent compared to 1990 levels during a second commitment period of the Protocol (2013–2020).

EU climate policy development has levelled off since the 2000s. The 2000s had seen a noticeable acceleration of climate policy development at the EU level, with the climate and energy package of 2008/09 marking a significant step for harmonization and communitarization of this policy area. In the wake of the financial and economic crises from 2008 onwards and the backlash of international climate policy at the 2009 Copenhagen climate summit, EU climate policy has become less dynamic. The climate and energy package has certainly been further implemented, and several new legislative acts (including Regulation (EU) No 517/2014 on fluorinated GHGs, Regulation (EU) No 333/2014 on carbon dioxide emissions of passenger cars and the Energy Efficiency Directive 2012/27/EU) have updated and strengthened the existing climate policies. However, discussions on developing the cornerstones of the EU's climate policy – the EU ETS, renewables and effort-sharing among the EU member states – have generally progressed at snail's pace. One illustration of this trend is the long and heated debate about postponing the auctioning of a certain amount of emission allowances (known as 'backloading') as a modest short-term measure to address the oversupply in the EU ETS. This measure was eventually passed in 2013, but more structural solutions to the problem continued to face an uncertain fate (Marcu, 2012).

As of 2014, the EU is nevertheless on its way to achieving a 20 per cent reduction in GHG emissions by 2020, and it will most likely achieve its renewable energy goal also. While the EU-15 (that has an 8 per cent reduction target under the Kyoto Protocol for 2008–2012) had achieved reductions of slightly over 15 per cent by 2012, the EU-28 had already

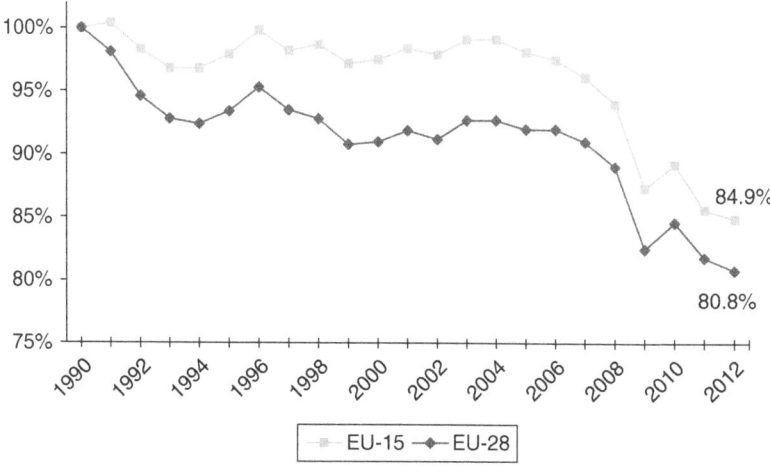

Figure 1.1 GHG emissions in the EU-28 and EU-15 between 1990 and 2012
Source: European Environment Agency GHG data viewer (eea.europa.eu, date accessed 31 July 2014).

achieved GHG emission reduction of over 19 per cent (see Figures 1.1 and 1.2). At the same time, GHG emissions in the EU-28 could decrease by more than 24 per cent by 2020, in excess of the 20 per cent reduction target (EEA, 2014, p. 42). The energy efficiency goal can also be achieved, but further implementing measures to 2020 are likely to be required (see Figure 1.2; EEA, 2014, p. 75; European Commission, 2014a, p. 4).

In January 2014, the European Commission proposed a new policy framework for climate and energy policy to 2030, as an interim step to the 2050 objective (European Commission, 2014b). A single GHG emission reduction target of 40 per cent by 2030, compared to 1990, and an EU-wide objective for expanding the share of renewable energy to 27 per cent are to drive the required transition. In mid-2014, the Commission furthermore suggested a 30 per cent target for energy efficiency for 2030. In October 2014, the European Council agreed to adopt the 40 per cent GHG emission reduction target, a target of expanding the EU's share of renewable energy to at least 27 per cent and a non-binding energy efficiency target of 27 per cent (European Council, 2014).

While any increase in ambition constitutes a step in the right direction, whether the proposed targets are sufficiently ambitious to move the EU towards decarbonization is a matter of debate (see Figure 1.3). Structural change needs to be initiated early on, since delays will make later changes more difficult (Neuhoff et al., 2014). In this respect,

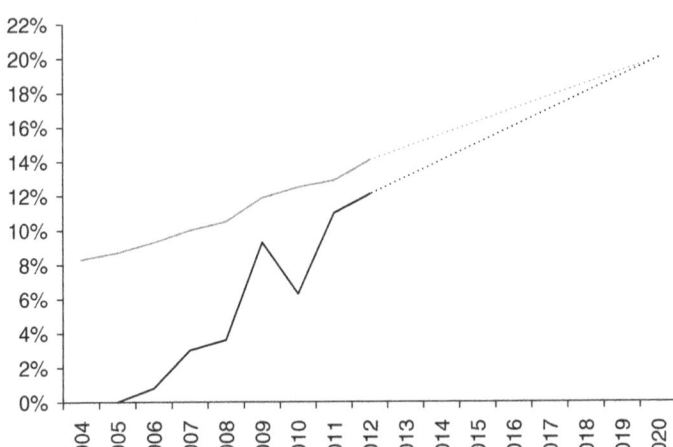

Figure 1.2   Progress towards 2020 renewable and energy efficiency targets in the EU-28

Note: Actual figures from 2004 (renewables) and 2005 (energy efficiency) to 2012. Share of renewable energy measured as a percentage share of final energy consumption per year. Energy efficiency improvements measured against baseline of 2005 and the business-as-usual forecast for final energy consumption to 2020. Linear trajectories to 2020 targets from 2012.
Source: Eurostat data, ec.europa.eu/Eurostat, date accessed 18 November 2014.

especially the targets for renewable energy and energy efficiency to 2030 have been criticised as insufficient.[1] At the same time, the EU's 2030 targets still are nominally the most far-reaching among the major international players.

Climate policy development in the EU since the late 2000s has occurred against the backdrop of a much changed and changing internal and external political context. The financial and economic crises from 2008 onwards have weakened the political priority accorded to climate policy, as have the political crises, such as in the Ukraine and the Middle East (Syria, Iraq, Palestine), and the Ebola crisis in Western Africa in 2014. The growing internal divergence of preferences, as most clearly visible from consistent Polish opposition to EU climate policy initiatives, was increasingly manifested as a major impediment to EU policy development. In contrast, by 2014, the Paris Climate Summit, scheduled for the end of 2015, where the conclusion of a new international climate agreement is expected, started to cast its shadow and arguably contributed to the agreement reached within the European Council in October 2014 (see above). Different to the Copenhagen Summit in 2009,

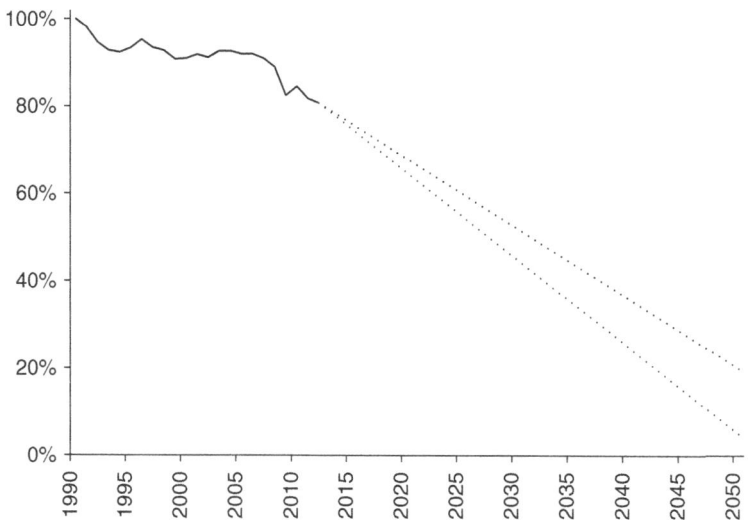

Figure 1.3   EU-28 GHG emission trajectory to reach reduction of 80–95 per cent by 2050

Note: Actual figures to 2012, linear trajectories to 2050.

Source: Own calculations based on EEA GHG data viewer (www.eea.europa.eu, date accessed 10 November 2014).

the EU has not worked towards translating its targets into EU law prior to the 2015 Paris conference.

2. To 2050: Long-term policy planning

Soon after the adoption of the climate and energy package for 2020, the European Council, in October 2009, stipulated an 'EU objective ... to reduce emissions by 80–95% by 2050 compared to 1990 levels' (European Council, 2009, para. 7). This long-term vision remains a political objective, without binding policy measures to implement it. It can also be argued that it does not represent a consensual and unconditional commitment and/or that it is politically over-ambitious or even unrealistic (Geden & Fischer, 2014). However, it is and remains in line with scientific estimates of the effort required to combat climate change. Even without consensual political blessing, the 2050 decarbonization objective provides a suitable benchmark for assessing the state and progress of EU climate policy.

On the basis of the aforementioned European Council conclusions, we furthermore assume, for the analysis in this volume, that there is a high level of political commitment to decarbonization in the EU.[2] This assumption seems to be in line with a number of roadmaps published by the European Commission in 2011 laying out pathways for achieving the 2050 decarbonization objective, including a 'Roadmap for Moving to a Competitive Low Carbon Economy in 2050' (European Commission, 2011b), the 'Energy Roadmap 2050' (European Commission, 2011a) and the 'Roadmap to a Single European Transport Area – Towards a Competitive and Resource Efficient Transport System' (European Commission, 2011d). To achieve the overarching 2050 goal of reducing GHG emissions in the EU by 80–95 per cent, the energy sector is expected to reduce its emissions to close to zero by 2050, and the transport sector by at least 60 per cent.

Besides the European Commission's roadmaps, various other research institutions and organizations have developed their own visions of how to reach the 2050 objective (see, for example, ECF, 2010; EREC & Greenpeace, 2010; EREC, 2010; EWEA, 2011; Heaps et al., 2009; PricewaterhouseCoopers, 2010; Shell International BV, 2008; WWF, 2011). While the various scenarios and roadmaps have differing assumptions and starting points, they aim to achieve at least an 80 per cent reduction in GHG emissions by 2050. The diversity among the scenarios stems from the choice of technology used to achieve the objective and from the various cost and public acceptance estimates inherent in these choices.

Carrying out a clear comparison among the many scenarios and roadmaps available is complex. The diversity of assumptions, aims and starting points of the scenarios means a comprehensive comparison based on clear points of reference is all but impossible. While some scenarios focus on the energy sector, others combine transport, agriculture, buildings and industry. Some scenarios explicitly provide absolute figures for their findings, but others rely on percentage point improvements. While some scenarios clarify their scenario-building starting points (for example, with regard to costs, technological developments, societal support), others do not. Therefore, we do not attempt to provide an in-depth analysis here of each of the scenarios.

Nevertheless, a review of these and the European Commission's own scenarios does point to a number of overarching conclusions about the road to 2050. First, there is no clear or 'correct' pathway to decarbonization by 2050. The many scenarios outline a range of possible actions that each lead to an ambitious reduction of GHG emissions by 2050. A combination of policy measures is likely to be required, with no single 'silver bullet' solution highlighted. Additionally, action is required across a number of policy sectors – including power, industry, transport, buildings and agriculture. However, the potential for deep reductions in GHG emissions among these sectors varies, with the power sector having the highest potential for reducing GHG emissions and the agricultural sector facing structural barriers that make the reduction of emissions particularly challenging (European Commission, 2011b, p. 5).

Second, two elements are highlighted as important for decarbonization across the scenarios: namely, improved energy efficiency and increased shares of renewable energy. These elements address the energy supply side (renewables – mostly power generation) and the demand side (energy efficiency – transport, heating, electricity demand). In all decarbonization scenarios of the European Commission, both energy efficiency and the share of renewables increase (European Commission, 2011a). Some stakeholders' scenarios outline that a fully renewable energy system is possible by 2050, and that such a strategy could achieve GHG emission reductions of more than 90 per cent (EREC & Greenpeace, 2010). Other scenarios point to higher levels of nuclear and CCS technology along with continued fossil-fuel use to counterbalance uncertainties about the ability of the electricity grid to handle large shares of variable renewable electricity (mainly from wind and solar power, see also Chapter 4) (ECF, 2010; Eurelectric, 2010). Nevertheless, even in scenarios where renewable energy and energy efficiency play less significant roles in 2050, the share of renewable energy increases. Ranges between 31 per cent (Eurelectric, 2010) and 92 per cent (EREC & Greenpeace, 2010; EREC, 2010) shares of renewable energy can be found in the scenarios, for example, and energy efficiency improves significantly to 2050. Thus, although improvement is required, scenarios differ as to the degree of renewable energy increase and energy efficiency improvements.

Third, the scope for the continued use of fossil fuels depends on the availability and commercial viability of CCS technology. CCS technology faces several obstacles in its development, including public acceptance, funding and commercialization. The longer it takes for CCS technology to become commercially viable, the less of a role it can play in moving to decarbonization. Moreover, for CCS to play a transformational role, significant pipeline and storage infrastructure projects would need to be implemented sooner rather than later (Fischedick et al., 2012; Förster et al., 2012). If CCS technology were not to become commercially viable, no new fossil-fuel power plants could be added to the EU's energy system. As these fossil-fuel power plants have lifetimes of about 50 years, any new plants added since

2000 will likely still be operational in 2050, leading several scenarios to dismiss all new fossil-fuel power plants for decarbonization (EREC & Greenpeace, 2010; Heaps et al., 2009). Given the delay the development and deployment of CCS technology has faced in Europe, suggestions exist that CCS technology cannot be considered a major element of any transition to decarbonization (Reichardt et al., 2012) and high-CCS scenarios become increasingly unlikely. Thus, while fossil fuels may indeed remain a part of the energy mix to 2050, this share will have to be reduced dramatically to achieve decarbonization, and the remaining small share will require (problematic) CCS technology to continue. However, the European Commission also highlighted, in its 2014 proposal for a climate and energy policy framework to 2030, the importance of CCS for GHG emission reductions in industrial processes, where only limited fuel switching and efficiency improvements can conceivably be achieved (European Commission, 2014b, p. 16; see also Chapter 5).

Fourth, the scenarios indicate the uncertain future of nuclear energy in the EU, despite its status as a 'low-carbon' energy source. In the Commission's energy sector decarbonization scenarios, the absolute amount of nuclear energy in gross inland consumption varies between 29 and 217 mega tonnes of oil equivalent (Mtoe) in 2050 (decreasing from 235 Mtoe in 2010). The extent of the decrease depends on public acceptance of nuclear energy, the increase in renewable energy, improvements in energy efficiency and the potential roll-out of CCS technology (European Commission, 2011c, pp. 56–77). Other scenarios rule out any role for nuclear energy in 2050, while managing to ensure decarbonization through heightened energy efficiency measures and a strong roll-out of renewable energy (ECF, 2010; EREC & Greenpeace, 2010; Heaps et al., 2009).

Fifth, many scenarios estimate that the costs of decarbonization will either be lower than or will not greatly exceed the increasing costs of a fossil-fuel-based economy into the future (depending on assumptions of GDP growth and technological commercialization) (Ackerman et al., 2009; Heaps et al., 2009; Hübler & Löschel, 2013; Stern, 2007). Such studies point to the rationality of taking long-term plans into consideration in short-term policymaking. Delaying action for short-term gains in times of economic constraint may in the long run not necessarily lead to cost savings, but to cost increases (Stern, 2007). Rather than consider costs, therefore, this book focuses on the policy requirements to bring about decarbonization.

Going beyond the scenarios, two further points deserve attention upfront. First, political action is required sooner rather than later to achieve decarbonization and to minimize the costs/maximize the benefits of doing so (Ackerman et al., 2009; Stern, 2007). The EU's agreed 2030 climate and energy policy framework does include an increase in the level of ambition as compared to 2020 to reduce GHG emissions by at least 40 per cent. Nevertheless, the way to achieve this target was not specified, with a renewable energy share increase from 20 to at least 27 per cent by 2030 and a non-binding energy efficiency goal of 27 per cent (European Council, 2014). Also according to the European Commission, long-term planning is required to spur early action and to ensure investment cycles in the energy sector align with goals for 2050 – otherwise there is a risk of locking in fossil-fuel or carbon-rich infrastructure (European Commission, 2010, p. 5). Infrastructure planning and development is one of the major keys to ensuring a successful transition to decarbonization, and planning over long-time horizons is crucial to make decisions that are compatible with the decarbonization agenda.

Second, strong international action on climate change would further alleviate the economic and political costs for the EU moving towards decarbonization by 2050. Fears about the EU going it alone, or losing out in terms of industry and economy to countries without similarly stringent climate objectives, may hold politicians back from taking the long-term decisions required (Babiker, 2005; European Commission, 2011b, pp. 13–14; Raihani & Aitken, 2011). It is therefore in the EU's own interest to push for international action on climate change, in both its bilateral and multilateral external relations, including in international climate negotiations under the UNFCCC (European Commission, 2014b, pp. 16–18).

In sum, despite the differences among the various scenarios, all confirm that further measures are required for decarbonization to become reality by 2050 – including far-reaching action in the near and medium term. Action needs to take place at multiple levels of governance, sooner rather than later, and across many policy sectors with a view to achieving the 2050 decarbonization goal. Policies to ensure a strong increase of renewable energies and energy efficiency are indispensable, while the contributions of nuclear energy and CCS remain uncertain and aggregate costs are low or at least manageable.

3. Analytical framework

The research and analysis of this book centres around one main question: What does decarbonization by 2050 mean for the EU's internal (energy-related) policies and external energy relations?

Accordingly, the volume takes the aforementioned objective of decarbonizing the EU by 2050 as its central point of reference. It engages with discussion on the EU's 2050 decarbonization ambition by examining the progress already made internally, the challenges remaining and the external implications of moving to a decarbonized EU. The authors bring long-term perspectives and objectives into policy choices and political relations in the short to medium term.

Given that the book analyses internal EU policies and also investigates EU external energy relations, different approaches are applied. The questions to ask with respect to the development of internal energy-related policies are different to those posed when it comes to analysing external relations. Internally, we are interested in

(1) analysing and understanding the progress made so far and
(2) the potential for further progress to achieve the decarbonization goal in the future.

Externally, the question is about the possible and/or likely impacts of EU decarbonization on bilateral relations and foreign policy.

With regard to *internal energy-related policies*, we aim to take stock of the progress towards decarbonization so far, investigate which sectoral policies would be required at the EU level to bring about decarbonization and shed light on the conditions underpinning sectoral policy development. We thus focus on three main sub-questions:

1. To what extent do existing policies in the relevant sector help achieve decarbonization by 2050?
2. What policy measures would be required in this sector to ensure the transition to decarbonization is achieved?
3. What are the internal sectoral drivers and barriers to achieving decarbonization and the full integration of this long-term decarbonization objective into sectoral policymaking?

Responding to the first two questions requires policy analysis and an assessment of progress so far, which involves comparing business-as-usual projections with the reality of policy measures, thus identifying the gaps to 2050. The third question, in contrast, requires us to identify a number of key factors upfront that can be assumed to serve as drivers and barriers. Since political, institutional and societal aspects will interact on the road to decarbonization to 2050 – given the dramatic economic and societal transformation required to achieve the 2050 objective of reducing GHG emissions by 80–95 per cent – the identified factors aim to go beyond sectoral and technical drivers and barriers, or opportunities and challenges, to consider also the functional, political, societal and institutional aspects. These all have clear roots in relevant theoretical approaches (as indicated below). We therefore analyse functional relations, political will or commitment, societal backing and the institutional set-up as important factors for understanding progress and potential for progress towards decarbonization.

The *functional relations* between a particular sectoral policy and decarbonization objectives may more or less support or hinder progress towards decarbonization. In other words, sectoral and decarbonization objectives may be in more or less synergy or conflict with each other. For example, policies to promote renewable energy may be more easily compatible with decarbonization than policies to ensure competitive energy prices in the EU. While the notion that policies 'spill over' into other policy areas has links to functionalist and neofunctionalist theories (Haas, 1958; Lindberg, 1963; Rosamond, 2005), we are particularly interested in whether or to what extent functional interrelations with the long-term decarbonization agenda have been recognized, seriously considered and taken up in concrete sectoral policy development. Policymakers' recognition of opportunities to advance win-win policies, or their recognition of the need for trade-offs in conflictual functional relations, can lay a path for policy development and coherence (Dupont & Oberthür, 2012). The analysis of functional interrelations should also allow us to identify potential for enhancing synergy in the future.

A major political aspect of moving to decarbonization includes a level of *commitment* from political leaders to achieve the objective and push for adequate sectoral policies. If policymakers recognize the functional interrelations, do political authorities support moves towards decarbonization in the sector? Achieving decarbonization means integrating long-term climate policy objectives into short- and medium-term sectoral policy development. Political commitment to such policy integration and to long-term policy planning is considered an important factor for the success of achieving cross-sectoral objectives like decarbonization (Lenschow, 2010; Sprinz, 2009). Investigating the level of such political commitment to decarbonization in the policy sectors discussed thus promises to enhance our understanding of achievements and existing opportunities or barriers.

In democratic societies, furthermore, general *societal backing* for a particular objective is important for the prospects of it being achieved. Political actors are unlikely to push for dramatic societal transformations in the absence of some level of civil society and citizen backing. Such societal backing can act as a driver for policy change, while its absence may present a barrier to ambitious policy action. Public support at European level can be measured through Eurobarometer surveys and citizen support in elections, but civil society movements can also reflect wider societal concerns. In the case of decarbonization, the inclusion of climate policy advocates in the policy negotiations can provide a further push for sufficient levels of policy ambition. Throughout this volume, the role, if any, played by civil society and citizens as well as climate policy advocates in pushing or blocking decarbonization through sectoral policies is therefore investigated (Bernauer & Gampfer, 2013; Kohler-Koch, 2010).

Finally, the *institutional set-up* in each of the policy sectors may also help advance understanding about the extent of progress towards decarbonization in each sector and the conditions for further progress in the future. When it comes to policymaking, institutions matter, and the way they function can influence the final policy output (Pierson, 1998). For example, whether competence on a particular policy lies more at EU or at national level may be relevant, since 'lack of EU competence' constitutes a powerful argument against EU-level action. Decision-making rules in each policy area may also push or block policy measures. Where one party holds a veto, it may be more difficult to agree on long-term, ambitious policy measures to achieve decarbonization in a particular policy area (Widgrén, 2009).

Turning to *the EU's external energy relations*, the perspective on the role of decarbonization for analytical purposes is different. Whereas decarbonization is primarily a policy objective (dependent variable) when it comes to internal policies, we take it as a given (independent variable) for analysing the EU's external energy relations. The impact decarbonization, including the phase-out of fossil-fuel imports, is likely to have on the EU's external energy relations is the focus. Accordingly, the questions investigated in this respect are:

1. What are the consequences for EU external energy relations that arise from a dramatic shift towards decarbonization internally within the EU? In particular, how will or could decarbonization affect the geopolitics of energy and the EU's relations with major (fossil-fuel) energy partners?
2. How can the EU and its external energy partners manage their evolving relations under decarbonization so as to mitigate any negative impacts and enhance synergy?

The aim with respect to the EU's external energy relations is thus not primarily to explain the development so far and identify the driving forces and barriers. Instead, the focus is on investigating where decarbonization of the EU could be driving these external relations in view of a number of other factors. A number of context- and case-specific factors are likely to shape future relations and thus drive relations along with decarbonization. These factors include the context of historical relations, available technical solutions, political and geopolitical possibilities. Contributing authors analysing EU external energy relations pay systematic attention to certain categories of factors. The EU's internal view of the partner and the recognition of long-term potential changes to the relationship due to decarbonization will inform whether or not decarbonization is taken seriously in external relations. The similarity or differences between the regulatory frameworks of the EU and the partner country and the possibilities for investment (especially in infrastructure) across borders are important identified elements. Furthermore, any institutional connections between the EU and its partners, beyond high-level political interactions, may provide a forum for exchange on opportunities to move to decarbonization.

Overall, the analysis of this volume is thus framed so as to provide an overarching view of the state of the art of EU internal policies on the road to decarbonization and the evolution of external energy relations in the context of decarbonization. A focus on politics, institutions and society allows us to investigate the prospects for achieving decarbonization internally, while the analyses of EU external energy relations aim to assess the likely impact of decarbonization, and the scope for managing this impact, in light of prominent contextual factors.

4. Book structure and chapter overview

Given its two-fold focus, the book is divided into two main parts. Six chapters (Chapters 2 to 7) discuss separate internal policy sectors that will play a central role in achieving decarbonization. Subsequently, four chapters (Chapters 8 to 11) turn the focus to external EU energy relations and especially the implications of decarbonization for traditional fossil-fuel energy partners. A concluding chapter brings together the main findings of the volume and draws some overarching lessons regarding the challenges, opportunities and implications of decarbonization for internal EU policymaking and for EU external energy relations.

Chapter 2 by Radostina Primova on the EU's internal energy market provides essential background information for the consideration of policy developments towards decarbonization. Primova describes the development of the internal energy market with an emphasis on the electricity market and investigates the interlinkages between the development of this market and moves to decarbonize by 2050. Primova argues that a fully competitive and functioning energy market is an essential condition for achieving EU decarbonization objectives. As this link is not always clear and evident, it has only slowly, but still insufficiently, been recognized in policymaking. As a result, synergies between the energy market and decarbonization objectives – which have been realized only to some extent due to their functional overlap – have not been sufficiently exploited; this is also a result of limited involvement of stakeholders, lack of political will, coordination and coherence. Primova identifies particular potential for future improvement with a strong focus on energy infrastructure development, which should be better aligned with EU decarbonization objectives, and by strengthening regional governance structures. Overall, she concludes that the link between decarbonization objectives and the internal energy market could be more clearly articulated by policymakers – for the benefit of both policy areas.

In Chapter 3, Stefan Lechtenböhmer and Sascha Samadi address the power sector that is at the front and centre of any efforts to decarbonize. They analyse the wealth of power sector decarbonization scenarios available and, on that basis, identify four key strategies for decarbonizing electricity generation and maximizing the use of low-carbon electricity for decarbonizing other energy uses in the EU:

(1) reducing demand (energy efficiency);
(2) replacing fossil fuels with electricity (for example, in heating and transport);
(3) increasing low-carbon electricity generation, especially renewables; and
(4) integrating increasing shares of fluctuating electricity generation from renewables (especially wind and solar) into the electricity grid.

The authors highlight progress already made to reduce GHG emissions in the EU's power sector, especially regarding increased renewables production, but also point to the great gap remaining to reach the decarbonization goal by 2050. Challenges remain in all four aforementioned areas on the demand and supply sides. While limits of competence constrain EU-level policy development in this field, the authors highlight the co-benefits of decarbonizing power production, including for energy security and employment. Such co-benefits may be particularly important in a political environment in which priorities have shifted and public support for decarbonization is tempered by employment concerns, for example.

Thomas Sattich adds further to the story of the power sector in Chapter 4, by focusing particularly on the EU's electricity infrastructure and the policies that are required to ensure the grid is updated for a decarbonized world. Sattich highlights the major technical challenges facing the EU in updating its grid to cater for higher volumes of variable renewably generated electricity. Solutions include promoting an EU-wide grid vision, providing targeted financial support for infrastructure development and enhancing research into energy storage and so-called smart grids. Sattich examines how functional overlap, political will, societal backing and institutional set-up have affected the extent of grid policy development for decarbonization at the EU level. He describes how grid development does not always follow a decarbonization perspective, especially when new fossil-fuel infrastructure gains policy support, and that the transition to a renewables-based system creates winners and losers. Political will has been hampered by fears of competition and security of member states, which is closely linked to the sensitive issue of the division of competences between the EU and its member states in this policy field. Policymaking on grid development takes place on multiple levels of governance, and coordination among the key stakeholders is not always easy. Furthermore, climate policy advocates are not necessarily considered among the key stakeholders. Sattich concludes that more policy measures and more efforts at coordination, across multiple levels of governance and among a wide variety of stakeholders, are required to develop the grid for decarbonization.

In Chapter 5, Max Åhman and Lars Nilsson discuss the implications of decarbonization for the industry sector in the EU, which accounts for about 20 per cent of EU CO_2 emissions. They focus, in particular, on the challenges of the energy-intensive industry (steel, cement, organic chemicals, etc.). The potential for emission reductions through incremental changes is limited so that decarbonization is likely to require new 'breakthrough technologies'. In this context, Åhman and Nilsson highlight the potential of biomass as a fuel and feedstock, CCS and electrification. The latter would have important repercussions for the overall energy system because of the implied significant increase in demand for electricity. Highlighting the importance of paying attention to trade and industrial competitiveness, the authors stress the need for a strengthened policy framework to support and push the timely development and demonstration of innovative breakthrough technologies. As development and demonstration is time-consuming and heavy industry's investment decisions have long-term impacts, action is required now to enable a shift to full decarbonization by 2050. Agreeing and implementing the required policies is hampered by the limited economic co-benefits, political differences within the EU and limited focus of societal interest groups on the challenges heavy industry faces in moving to decarbonization.

In Chapter 6, Tom van Lier and Cathy Macharis take up the discussion of decarbonization in the EU's transport sector, responsible for about a quarter of the EU's GHG emissions. Past efforts to limit GHG emissions from transport, including a range of policies at the EU level, have been insufficient for putting emissions on a trajectory consistent with ultimate emission reductions of 60–80 per cent from 1990 levels by 2050 in the transport sector. To achieve these reduction goals, a range of technical and non-technical solutions is required. Policies need to simultaneously improve:

(1) the GHG intensity of energy used by the transport sector;
(2) the efficiency of transport vehicles by both technical and operational means; and
(3) the efficiency of the transport system.

The authors highlight the need to enhance political will and broad stakeholder involvement in policymaking as well as the promise of more coherent emission reduction policies in transport through better cooperation across different levels of governance. This may enable better exploitation of the potential for synergy between decarbonization and transport policy objectives, while possible trade-offs also require attention and analysis.

The discussion on the buildings sector in Chapter 7 closes the internal EU policy sector analysis. In this chapter, Elin Lerum Boasson and Claire Dupont analyse EU policies to improve the energy performance of buildings and outline the policy gaps remaining for decarbonization in this sector. They demonstrate that the effectiveness of EU policies in this field has remained constrained despite their growth. This may be partially due to a lack of engagement of climate policy advocates, both in industry (which is rather decentralized in this field) and among

NGOs (which face resource constraints). It may also be traced back to the fact that energy efficiency in buildings is hardly an end in itself but is being pursued for other purposes, including combating climate change and enhancing energy security. The main barrier to progress in this field at the EU level seems, however, to be related to competence: policy development and implementation have consistently been delayed or diluted due to subsidiarity concerns. This raises questions about the appropriate and most effective role of EU policies in this field. One way forward may be the increased use of incentives, including different funding opportunities that may 'leverage' action within member states. Exploiting the windows of opportunity opened by external events (such as energy security fears) may help push political commitment to improve the energy performance of buildings. Skilled and dedicated policy entrepreneurs may be required to push such an agenda. More research into the multi-level governance characteristics of this policy field may be required to find further opportunities, especially with regard to cross-border cooperation in this field within the EU.

Chapter 8 by Tom Casier opens the analysis of the implications of the EU's decarbonization agenda on EU external energy relations. Casier discusses the overall geopolitics of the EU's decarbonization strategy. He argues that, under decarbonization, EU external energy relations will no longer depend on the control over resources and infrastructure. Rather, the geopolitical effects will depend on perceptions, frames and the development of regional and global energy markets in a wider context. While long-term projections are thus difficult to make, Casier expects that decarbonization would leave some room for gas imports while also potentially creating demand for renewable electricity imports. As a result, Russia, Norway and Algeria may remain prominent because of their gas reserves and could possibly balance losses with respect to oil and gas by developing their renewables potential. The Caspian Sea region, Turkey and Ukraine are likely to remain or become important partners with respect to gas imports, while the Persian Gulf states may lose importance as EU energy partners. Northern Africa could acquire additional significance with respect to renewable electricity. Casier concludes that the energy relations of a decarbonized EU will become more regionalized and will depend greatly on the development of infrastructure, technology and regulatory arrangements (rather than on control over resources).

In Chapter 9, Claire Dupont discusses EU energy relations with the Caspian Sea region in more detail. She focuses on relations with Azerbaijan and Turkmenistan, as potential suppliers of natural gas for the EU. Dupont examines the role of natural gas in the EU energy mix today and in the future under decarbonization. She questions whether the EU demonstrates a consistent logic in promoting natural gas infrastructure to connect to potential supplies in the Caspian region while simultaneously pushing for decarbonization internally. Without widespread implementation of CCS technology, further expansion of fossil-fuel resources in the EU cannot be supported under decarbonization in the medium to long term. Furthermore, relations with the Caspian region are situated in the broader political context of the region, including Russian and Chinese interests there. EU relations with the region are based not only on energy, but also on EU ambitions to promote democracy and human rights in its neighbourhood. Dupont explores how decarbonization could allow the EU to place more emphasis on democratization in its relations with Azerbaijan and Turkmenistan, but also highlights that the broader political context must be carefully considered. She concludes that placing decarbonization at the heart of external relations presents the EU with the opportunity to develop a long-term strategy in fashioning its relations with the Caspian region.

Olga Khrushcheva and Tomas Maltby tackle the evolution in EU–Russia energy relations under decarbonization in Chapter 10. The EU has historically been reliant on Russia for a large portion of its fossil energy supplies. As the EU moves towards decarbonization, how will the lower demand for fossil fuels impact relations with Russia? Are there options for managing, maintaining or improving these relations while moving towards decarbonization? Khrushcheva and Maltby describe the broader tensions and challenges in EU–Russia energy relations (especially in the context of the Ukraine crisis since 2013 and previous gas crises in 2006 and 2009), while also pointing out the opportunities for Russia to remain an energy partner of the EU through renewable energy and energy efficiency measures. The authors also describe some initial steps, politically and institutionally, that are being taken along this road. They conclude that several challenges remain in EU–Russia energy relations more broadly but that creative solutions can be found and co-benefits just need to be exploited.

In Chapter 11, Torbjørg Jevnaker, Leiv Lunde and Jon Birger Skjærseth take us through the potential for change in EU–Norway energy relations. Norway is a reliable partner of the EU and a major supplier of, especially, natural gas. It also is partly integrated into the EU regulatory framework through the European Economic Area. Nevertheless, the authors argue that engaging Norway in decarbonization comes up against the country's fossil-fuel interests (oil and gas) that are likely to remain strong for the foreseeable future. They consider three scenarios for the future development of EU–Norway relations including one in which the EU does realize its decarbonization aspiration bolstered by strong international climate cooperation. Only in this case, the authors find it likely that Norway's petroleum interests would be seriously challenged. EU decarbonization would thus pose the challenge of replacing petroleum trade with something else in the bilateral relationship. Both renewable energy and electricity trade, with Norway acting as a 'green battery' for the EU storing excess intermittent renewable energy production, provide interesting avenues, but currently do not seem to have the potential of an equivalent replacement. Despite the favourable starting point, decarbonization would thus constitute a considerable challenge for the development of EU–Norway relations.

Finally, in Chapter 12, we present some overarching conclusions from the discussions and analysis of earlier chapters. Although each chapter highlights several contextual conclusions, a number of broader lessons and conclusions can be drawn. Among the lessons we discuss in Chapter 12, we highlight that decarbonization is not only a challenge to be overcome, but also presents many opportunities for both internal policies and external strategies. We also underscore the importance of strong political will to ensure that decarbonization is achieved, which can also aid in maintaining, and even improving external relations. Furthermore, the gaps and limits in internal policies and external energy strategies from a decarbonization perspective can be understood as products of political systems and institutions and short-term policy visions. Breaking free of this mould will require considerable political courage. External shocks may also move policy further. Technical solutions already exist, and more are being developed and come on stream as we write. Further technological breakthroughs may ease the transition to decarbonization. Finally, throughout the chapters, it becomes clear that integrating better long-term planning for decarbonization into policy development and external relations is required.

Notes

1. See   http://www.euractiv.com/sections/eu-priorities-2020/eu-leaders-adopt-flexible-energy-and-climate-targets-2030-309462 for early critical comments in response to the European Council's conclusions, date accessed 5 November 2014.
2. Please note, however, that much of the analysis would hold even without this assumption.

References

Ackerman, F., Stanton, E. A., Decanio, S. J., Goodstein, E., Howarth, R. B., Norgaard, R. B., Norman, C. S. & Sheeran, K. A. (2009) *The Economics of 350: The Benefits and Costs of Climate Stabilization* (Portland, OR: E3Network).

Adelle, C., Russel, D. & Pallemaerts, M. (2012) 'A "Coordinated" European Energy Policy? The Integration of EU Energy and Climate Change Policies', in F. Morata & I. Solorio Sandoval (eds.), *European Energy Policy: An Environmental Approach* (Cheltenham: Edward Elgar), pp. 25–47.

Babiker, M. H. (2005) 'Climate Change Policy, Market Structure, and Carbon Leakage', *Journal of International Economics, 65*, 421–445.

Bernauer, T. & Gampfer, R. (2013) 'Effects of Civil Society Involvement on Popular Legitimacy of Global Environmental Governance', *Global Environmental Change, 23*(2), 439–449.

Dupont, C. & Oberthür, S. (2012) 'Insufficient Climate Policy Integration in EU Energy Policy: The Importance of the Long-term Perspective', *Journal of Contemporary European Research, 8*(2), 228–247.

ECF (2010) *Roadmap 2050: A Practical Guide to a Prosperous, Low Carbon Europe* (Brussels: European Climate Foundation).

EEA (2014) 'Trends and Projections in Europe 2014: Tracking Progress towards Europe's Climate and Energy Targets for 2020', *EEA Report*, No. 6/2014 (Copenhagen: European Environment Agency).

EREC (2010) *RE-Thinking 2050: A 100% Renewable Energy Vision for the European Union* (Brussels: European Renewable Energy Council).

EREC & Greenpeace (2010) *Energy [R]evolution: Towards a Fully Renewable Energy Supply in the EU 27* (Brussels: Greenpeace International and European Renewable Energy Council).

Eurelectric (2010) *Power Choices: Pathways to Carbon-Neutral Electricity in Europe by 2050* (Brussels: Eurelectric).

European Commission (2010) 'Energy Infrastructure Priorities for 2020 and Beyond – A Blueprint for an Integrated European Energy Network', COM(2010) 677.

European Commission (2011a) 'Energy Roadmap 2050', COM(2011) 885/2.

European Commission (2011b) 'A Roadmap for Moving to a Competitive Low Carbon Economy in 2050', COM(2011) 112.

European Commission (2011c) 'Impact Assessment Accompanying the Document: Energy Roadmap 2050', SEC(2011) 1565 Part Two.

European Commission (2011d) 'White Paper – Roadmap to a Single European Transport Area. Towards a Competitive and Resource Efficient Transport System', COM(2011) 144.

European Commission (2014a) 'Energy Efficiency and Its Contribution to Energy Security and the 2030 Framework for Climate and Energy Policy', COM(2014) 520.

European Commission (2014b) 'A Policy Framework for Climate and Energy in the Period from 2020 to 2030', COM(2014) 15.

European Council (2009) 'Presidency Conclusions', Document 15265/1/09, 29 & 30 October.

European Council (2014) 'Conclusions', Document EUCO 169/14, 23 & 24 October.

EWEA (2011) *EU Energy Policy to 2050: Achieving 80–95% Emissions Reductions* (Brussels: European Wind Energy Association).

Fischedick, M., Förster, H., Friege, J., Healy, S., Lechtenböhmer, S., Loreck, C., Matthes, F. C., Prantner, M., Samadi, S. & Venjakob, J. (2012) *Power Sector Decarbonization: Metastudy. Final Report for the SEFEP Funded Project 11–01* (Berlin: Öko-Institut e.V. and Wuppertal Institute).

Förster, H., Healy, S., Loreck, C., Matthes, F., Fischedick, M., Samadi, S. & Venjakob, J. (2012) 'Information for Policy Makers 1: Decarbonization Scenarios Leading to the EU Energy Roadmap 2050', *SEFEP Working Paper, 23* (January).

Geden, O. & Fischer, S. (2014) 'Moving Targets: Negotiations on the EU's Energy and Climate Policy Objectives for the Post-2020 Period and Implications for the German Energy Transition', *SWP Research Paper*, March 2014 (Berlin: Stiftung Wissenschaft und Politik).

Haas, E. B. (1958) *The Uniting of Europe: Political, Social and Economic Forces, 1950–1957* (Stanford, CA: Stanford University Press).

Haug, C., Rayner, T., Jordan, A., Hildingsson, R., Stripple, J., Monni, S., Huitema, D., Massey, E., van Asselt, H. & Berkhout, F. (2010) 'Navigating the Dilemmas of Climate Policy in Europe: Evidence from Policy Evaluation Studies', *Climatic Change, 101*(3–4), 427–445.

Heaps, C., Erickson, P., Kartha, S. & Kemp-Benedict, E. (2009) *Europe's Share of the Climate Challenge: Domestic Actions and International Obligations to Protect the Planet* (Stockholm: Stockholm Environment Institute).

Hübler, M. & Löschel, A. (2013) 'The EU Decarbonization Roadmap 2050 – What Way to Walk?', *Energy Policy, 55*, 190–207.

IPCC (2007) *Climate Change 2007. Fourth Assessment Report: Synthesis Report* (Geneva: Intergovernmental Panel on Climate Change).

IPCC (2013) 'Summary for Policymakers', in T. F. Stoker, D. Qin, G.-K. Plattner, M. Tignor, S. K. Allen, J. Boschung, A. Nauels, Y. Xia, V. Bex & P.M. Midgley (eds.), *Climate Change 2013: The Physical Science Basis. Contribution of Working Group I to the Fifth Assessment Report of the Intergovernmental Panel on Climate Change* (Cambridge: Cambridge University Press).

Jordan, A., Huitema, D., van Asselt, H. & Rayner, T. (2010) 'The Evolution of Climate Change Policy in the European Union: A Synthesis', in A. Jordan, D. Huitema, H. van Asselt, T. Rayner & F. Berkhout (eds.), *Climate Change Policy in the European Union: Confronting the Dilemmas of Mitigation and Adaptation?* (Cambridge: Cambridge University Press), pp. 186–210.

Kohler-Koch, B. (2010). 'Civil Society and EU Democracy: "Astroturf" Representation?', *Journal of European Public Policy, 17*(1), 100–116.

Lenschow, A. (2010) 'Environmental Policy: Contending Dynamics of Policy Change', in H. Wallace, M. A. Pollack & A. R. Young (eds.), *Policy-Making in the European Union,* 6th edn. (Oxford: Oxford University Press), pp. 307–330.

Lindberg, L. N. (1963) *The Political Dynamics of European Economic Integration* (Palo Alto, CA: Stanford University Press).

Marcu, A. (2012) 'Backloading: A Necessary but Not Sufficient First Step', *CEPS Special Report,* 20 November (Brussels: Centre for European Policy Studies).

Neuhoff, K., Acworth, W., Dechezleprêtre, A., Dröge, S., Sartor, O., Sato, M., Schleicher, S. & Schopp, A. (2014) *Staying with the Leaders: Europe's Path to a Successful Low-Carbon Economy* (London: Climate Strategies).

Oberthür, S. & Pallemaerts, M. (eds.) (2010) *The New Climate Policies of the European Union: Internal Legislation and Climate Diplomacy* (Brussels: VUB Press).

Oberthür, S. & Roche Kelly, C. (2008) 'EU Leadership in International Climate Policy: Achievements and Challenges', *International Spectator*, 43(3), 35–50.

Pierson, P. (1998) 'The Path to European Integration: A Historical-Institutionalist Analysis', in W. Sandholtz & A. Stone Sweet (eds.), *European Integration and Supranational Governance* (Oxford: Oxford University Press), pp. 27–58.

PricewaterhouseCoopers (2010) *100% Renewable Electricity: A Roadmap to 2050 for Europe and North Africa* (UK: PricewaterhouseCoopers), www.pwc.com/climateready, date accessed 14 January 2012.

Raihani, N. & Aitken, D. (2011) 'Uncertainty, Rationality and Cooperation in the Context of Climate Change', *Climatic Change*, 108, 47–55.

Reichardt, K., Pfluger, B., Schleich, J. & Marth, H. (2012) 'With or Without CCS? Decarbonising the EU Power Sector', *Responses Policy Update*, 3 July 2012.

Roche Kelly, C., Oberthür, S. & Pallemaerts, M. (2010) 'Introduction', in S. Oberthür & M. Pallemaerts (eds.), *The New Climate Policies of the European Union: Internal Legislation and Climate Diplomacy* (Brussels: VUB Press), pp. 11–25.

Rosamond, B. (2005) 'The Uniting of Europe and the Foundation of EU Studies: Revisiting the Neofunctionalism of Ernst B. Haas', *Journal of European Public Policy*, 12(2), 237–254.

Schreurs, M. A. & Tiberghien, Y. (2007) 'Multi-level Reinforcement: Explaining European Union Leadership in Climate Change Mitigation', *Global Environmental Politics*, 7(4), 19–46.

Shell International BV (2008) *Shell Energy Scenarios to 2050* (The Hague: Shell International BV).

Sprinz, D. F. (2009) 'Long-term Environmental Policy: Definition, Knowledge, Future Research', *Global Environmental Politics*, 9(3), 1–8.

Stern, N. (2007) *The Economics of Climate Change: The Stern Review* (Cambridge: Cambridge University Press).

Widgrén, M. (2009) 'The Impact of Council Voting Rules on EU Decision-making', *CESifo Economic Studies*, 55, 30–56.

Wurzel, R. K. W. & Connelly, J. (eds.) (2011) *The European Union as a Leader in International Climate Change Politics* (London: Routledge).

WWF (2011) *The Energy Report: 100% Renewable Energy by 2050* (Gland, Switzerland: World Wide Fund for Nature).

2
The EU Internal Energy Market and Decarbonization

Radostina Primova

Introduction

This chapter focuses on the internal energy market development as an important tool to reach the decarbonization objectives agreed by the European Council in 2009 and anchored in the 2050 Energy Roadmap (see Chapter 1). The EU internal energy market constitutes the integration of EU member states' gas and electricity markets into one single market based on the free movement of goods, services, capital and persons. It aims to create free and fair competition in the energy sector, in which consumers (both industrial and household) can freely choose their supplier, and suppliers can provide gas and electricity without any restrictions across borders. Apart from this freedom of choice, fair competition would also require harmonized national regulations and market structures that are free from dominant players (Eikeland, 2011, p. 15). Since natural gas is a fossil fuel that emits greenhouse gases, it is not easily compatible with achieving EU long-term decarbonization objectives (Dupont & Oberthür, 2012; see also Chapter 9). For this reason, the chapter focuses on the EU electricity market and its role in attaining EU climate policy goals for 2050.

The link between the internal energy market and decarbonization has not always been direct and evident. However, a common electricity market could be one of the major tools for advancing EU climate change objectives rather than a goal in itself. The creation of a fully competitive and well-functioning electricity market is an essential condition for achieving EU climate policy objectives. First, increasing competition in the energy sector would remove bottlenecks on the EU energy market and allow independent producers of renewable sources of energy (RES) to have better access to and benefits from the common market.

Due to their intermittent nature, some RES require well-interconnected regional and European electricity markets. The fragmentation of electricity markets could thus endanger their adoption on a larger scale (Schuh et al., 2012). In this context, promoting competition in the electricity sector could challenge the dominant positions of vertically integrated energy monopolies and enable better access to the grid for independent producers of renewable energy. Non-discriminatory access to electricity networks is a necessary condition for ensuring competitiveness on the energy market.

The main hurdle in the liberalization process is that electricity transmission and distribution networks in Europe have historically been owned by vertically integrated companies that have built monopolies over generation, transmission, distribution and supply activities. These companies have numerous incentives and possibilities to discriminate their competitors as far as access to their transmission and distribution network is concerned (Cabau, 2010; see also Chapter 4). The completion of the internal energy market is thus an important prerequisite for integrating renewables into the power grid in order to promote coordination between energy market actors to ensure that renewably generated electricity is well connected (Schuh et al., 2012; Dupont & Primova, 2011).

Second, the competitiveness of the EU energy market also has direct implications for the development of modern energy infrastructure, which is necessary for the absorption of the large volumes of intermittent RES. Therefore, investments in infrastructure development and expansion are indispensible for creating a level playing field for renewables and breaking the fossil fuel dependency pattern (see Chapter 4).

Third, an integrated energy market has a high cost-saving and risk-management potential (ECF, 2013, p. 7). A number of scenarios have illustrated that cross-border resource sharing could lead to considerable cost savings (Newbery et al., 2013; ECF, 2011). A more interconnected energy market will increase effective market size, allow for higher system flexibility and resource optimization, enable the balancing of resources, pooled research and development and deployment costs and will improve the overall system efficiency – with positive impacts on decarbonization (ECF, 2013, pp. 15–16).

This chapter first tracks the historical development of the EU internal energy market from the perspective of decarbonization. It is clear that decarbonization is more or less important at different stages of policy development. Next, I explain certain loopholes in EU energy market development and the limited integration of climate policy in this policy field with the help of the four main variables introduced in Chapter 1 (functional overlap; political will; societal backing; and institutional set-up). Then, I identify some current challenges on the road to decarbonization from an energy market perspective and conclude with some possible solutions and policy recommendations to combat these challenges.

1. Historical background

The development of the EU's internal energy market can be divided into two main time periods: the early days from the 1980s to the development of the first and second energy market packages in the mid-1990s and early 2000s; and the agreement of the third energy market package in the late 2000s.

1.1 Early days: Leading to the first and second energy packages

In the early stages of European energy market integration, decarbonization played a minor role in policy and legal developments. The major goal of the first Gas and Electricity Directives in the 1990s (Directive 98/30/EC; Directive 96/92/EC) was to achieve free and fair competition in the energy sector by gradually opening up the gas and electricity markets in member states and thus providing benefits for European consumers.

The first energy market initiatives were agreed under the EU's competence on the internal market programme. Competition policy, environmental objectives and foreign policy competences were also highly relevant (Andersen, 2000). Since energy policy was not constituted as a separate policy domain in the treaties at this time, European institutions used legal competences concerning environmental policy and the internal market programme in the treaties to propose and pass new energy legislation (Buchan, 2010, p. 360). Although energy policy was generally affected by some common environmental objectives, no agreement existed as to what measures to apply to achieve these objectives (Andersen, 2000).

At this early stage, climate change concerns did not play a role in the deliberations on a common EU energy market. The roots of EU energy market policy could be traced back to the Single Market Initiative in 1985, which aimed at creating a free and competitive internal market and removing barriers to free movement. EU policymakers became aware that an integrated energy market could boost the effectiveness of the internal market by decreasing energy costs and improving security of supply (Johnston & Block, 2012, p. 13). Breaking up vertically integrated utilities dominating the domestic energy markets ('unbundling') was seen as paramount for achieving free and fair competition in the energy sector (Eikeland, 2011, p. 17). Unbundling refers to the effective separation of network assets, in particular referring to the separation of generation and supply activities from transmission and distribution activities, in order to prevent the discrimination of third-party access to the grid. Depending on the degree of separation, unbundling can take different forms – accounting unbundling, management unbundling, legal unbundling, independent transmission operator unbundling or ownership unbundling (Jones, 2010, pp. 10–11).

The decision-making process leading to the first package of Gas and Electricity Directives was lengthy and thorny. The first and second liberalization packages contributed only to a limited extent to a level playing field for renewables to compete with fossil fuels. The 1996 Electricity Directive (Directive 96/92/EC) contained a small part of the Commission's initial proposals and allowed only a limited number of high volume electricity consumers the right to freely choose their supplier. Instead of strict instructions, the Directive offered only a framework for future liberalizations, leaving it up to member states to decide which regulatory measures to adopt (Eikeland, 2004, p. 6).

The legal foundation of the internal energy market was strengthened with the second package of liberalization measures, adopted in June 2003. This package consisted of the second Gas and Electricity Directives as well as a Regulation on cross-border trade in electricity (Directive 2003/54/EC; Directive 2003/55/EC; Regulation (EC) No 1228/2003). The package enshrined the right of third parties to non-discriminatory access to transmission and distribution systems in the gas and electricity sectors. It also stipulated requirements for legal unbundling for energy transmission networks and set up regulatory authorities in member states for monitoring purposes. Legal unbundling refers to the separation of network business, in which a separate legal undertaking is set up that carries out the activities of the network company (Jones, 2010, p. 11). Legal unbundling is limited to the organizational separation of units operating generation and supply from those operating transmission (Eikeland, 2011). Furthermore, regulatory governance in the energy sector was strengthened through the establishment of the European Regulators Group of Electricity and Gas (ERGEG), which set up a framework for coordinating national energy regulators.

Regarding the progress of establishing an internal energy market, a study by Jamasb and Pollitt (2005) shows that the first and second packages achieved a degree of standardization of structures, institutions and rules in national electricity markets. Despite the transformation of the domestic energy markets, the implementation of this legislation remained very slow, due partly to protectionist pressures for maintaining national monopolies (Aalto & Westphal, 2008).

At this stage, the link between the EU internal energy market and environmental policy was not evident and not recognized during the policy process. The main drivers behind internal energy market initiatives were the internal market programme, EU competition rules and consumer and regulatory issues. However, the gradual harmonization of rules and practices was a step forward to a more competitive and better-regulated energy market, which contributed indirectly to EU climate policy objectives by facilitating access of renewables. Drawing on the interconnectedness between the competitiveness and sustainability goals, one can infer a functional overlap between both objectives.

1.2 The third energy package and its climate policy dimension: Late 2000s

The third phase of the EU internal energy market development marked more significant progress towards decarbonization. This was due to the more competitive conditions for RES producers and an enhanced regulatory framework. The improvement of regulatory conditions led to the removal of further barriers in electricity trade, thus creating more incentives for investment in new infrastructure (see also Chapter 4).

Energy regulatory measures prior to 2009 envisioned legal, functional and accounting unbundling for transmission system operators and distribution system operators. However, the sector inquiry launched in 2005 and the country reviews conducted by the Commission in 2006 showed that these measures were insufficient for removing the conflict of interests arising from vertical integration. Market distortions and barriers to free competition were still evident (European Commission, 2007b; 2007c). The results from the energy sector inquiry in 2007 exposed the high concentration of electricity markets at the wholesale level and vertical foreclosure. This led to a structural conflict of interests and to a negative impact on market functioning, and reduced incentives to invest in networks. Based on these findings, the Commission concluded that ownership unbundling is the most efficient way to tackle these shortcomings and promote investment (European Commission, 2007c, p. 14; 2007d, p. 7). Ownership unbundling contains stronger unbundling measures than legal unbundling. This model refers to the full separation of supply and transmission assets. In this case, the vertically integrated undertaking has to sell its network assets to shareholders that are not involved in the generation, production or sale activities of the company (Jones, 2010, p. 11).

Driven by these overarching objectives and growing political concerns, the EU adopted a third package of liberalization in 2009, which aimed to fill in gaps from the previous packages and tackle the defects addressed in the sector inquiry and the benchmarking reports. The package introduced further unbundling provisions to ensure the effective separation of production and transmission assets and increased regulatory powers of the EU in the energy policy field (Dupont & Primova, 2011). The third energy package allows the three models of separation of supply and generation from transmission activities to co-exist:

- Full ownership unbundling requiring a full separation between supply and transmission activities;
- Allowing vertically integrated companies to retain ownership of their network assets, but requiring the transmission network to be managed by an independent system operator; and
- Enabling vertically integrated companies to keep their ownership of the transmission system provided it is managed by an independent transmission operator and subject to a number of safeguard provisions.

(Directive 2009/72/EC; Directive 2009/73/EC)

The harmonization of national energy regulations is another important element of decarbonization. It will lead to the optimization of cross-border electricity trade and facilitate the integration of intermittent RES through the creation of common rules on cross-border electricity trade and transparency. In this respect, the third liberalization package enhanced the regulatory powers and independence of national energy regulators from companies and governments. These regulators oversee the application of the EU's energy market rules. The third package also created an EU Agency for the Cooperation of Energy Regulators (ACER). ACER's main task is to monitor and improve cross-border regulatory harmonization in the gas and electricity sectors. EU energy regulation is further strengthened by the 2009 Electricity and Gas regulations (Regulation (EC) No 714/2009; Regulation (EC) No 714/2009). These contain provisions for harmonizing rules for cross-border exchanges in electricity and gas and for the establishment of a new framework for cooperation of transmission system operators at EU level, namely a European Network of Transmission System Operators for Electricity (ENTSO-E) (Dupont & Primova, 2011). Such enhanced coordination and harmonization in the electricity sector led also to more favourable conditions for the integration of renewables into the power grid and reducing barriers in cross-border electricity trade.

The Commission's persistence in advancing the liberalization process in the energy sector and ensuring open access to energy networks reflected the conviction that a competitive and well interconnected energy market would contribute to tackling multiple challenges. These include energy security challenges – through the transfer of emergency stocks around the EU – and climate change – by optimizing efficient energy use and reducing greenhouse gas emissions by allowing fair competition for RES (Buchan, 2010, p. 362), so that they can compete on a level playing field with fossil fuel industries.

The June 2011 benchmarking report monitoring the progress achieved in the internal gas and electricity market in the period 2009–2010 highlighted, however, that the integration of the electricity and gas markets in Europe was insufficiently developed at retail level (European Commission, 2011c). The report pointed to the limited interconnection capacity between member states, the existence of bottlenecks and the lack of harmonized market rules as major barriers for achieving an integrated wholesale energy market. There remained several implementation deficits at this time.

However, although the third package itself may have faced some challenges, especially in implementation, parallel policy processes in climate and energy brought the decarbonization agenda more to the fore in internal energy market policy also. In 2009, the EU adopted its energy and climate package for 2020 (see Chapter 1). The European Council later also considered the energy roadmap 2050, in which the Commission analysed many sectors – power generation, transport, industry, agriculture (European Commission, 2011b) – and their potential contribution to decarbonization (see Chapter 1). During the policymaking process of the third energy package, EU climate policy objectives thus gained relevance since an integrated and well-functioning market is important for RES integration into the grid. Furthermore, the Commission (2011a) presented a legislative proposal for an energy infrastructure package. This package aims to promote the modernization and expansion of Europe's energy infrastructure and the interconnection of networks across borders in order to make them capable of absorbing the growing share of RES and to complete the internal market in the power sector (see Chapter 3). In its 2012 Communication 'Making the Internal Energy Market Work', the Commission emphasized the role of the market, which, in combination with EU Emissions Trading Scheme, could drive investments and deliver a high-quality electricity system (European Commission, 2012a).

While decarbonization has never been the single overarching motivation for the development of the internal energy market, it can be said that the decarbonization agenda is served by the liberalization of the internal energy market. In the later years of policy development, this link seems to have become more explicit in policymaking.

2. Major challenges for decarbonization in EU internal energy market

The major challenges for decarbonization of the EU internal energy market can be divided into five main categories: competence-related, infrastructural, regulatory, technological and socio-economic.

The first cluster of challenges relates to the division of competences in the field of RES policy. Given the policy differences in the promotion of renewables among member states (Reiche & Bechberger, 2004) and the different national scenarios for future energy mixes, the deployment of new energy technologies and the adaptation of their power and transport sectors will depend on member states. Since renewable energy policy instruments are considered a national matter, the harmonization of national support schemes is rather unlikely in the short term (Nilsson, 2011). Shifting more competence to the EU level in the field of renewable energy policies appears also unlikely, except for reasons of security of supply (Nilsson, 2011, p. 126). The lack of a truly integrated electricity market also has implications for the effectiveness of RES support schemes. Renewable energy stakeholders share the view that the harmonization of national support schemes is difficult to achieve without a fully integrated and well-functioning electricity market.[1]

Second, one of the major challenges in advancing the internal energy market project for the purpose of EU decarbonization objectives is the low investment in new infrastructure required to speed up the process of integrating renewables into the grid. The European Parliament shows that the fragmentation of energy markets could serve as a barrier to the adoption of RES because of the lack of incentives to build the necessary infrastructure (Schuh et al., 2012, p. 28).

This leads, third, to the regulatory challenges for decarbonization of the EU electricity markets. In most cases, investment decisions are taken on the basis of long-term expectations and price signals from electricity markets (Schuh et al., 2012, p. 29; Isabel & Soares, 2004). One of the biggest hurdles in infrastructure development has also been the lack of a stable regulatory framework in member states. Investors' hesitation has often been attributed to four main factors: low demand, short- and medium-term overcapacity, regulatory instability and deteriorating financial conditions (Capgemini, 2012). Creating a stable regulatory framework would be key to tackling these hurdles to decarbonization.

The asymmetry in climate regulation on a global level could also adversely affect EU decarbonization efforts. The lack of cost convergence between low-carbon and high-carbon energy systems as a result of global competition between low-carbon economies, such as the EU, and those not committed to significant emission reductions could undermine EU internal regulatory mechanisms to tackle climate change in the energy sector. This could be seen as a consequence of the lock-in of high-carbon assets (ECF, 2013, p. 15). Other political and regulatory challenges concern uncertainties about future climate policies, fuel prices, technology developments, and the EU energy and climate framework to 2030 and beyond (see Chapter 1; Thema et al., 2013, p. 15).

Fourth, the technological challenges to decarbonization are often linked to regulatory gaps in EU energy market legislation. The transition to a low-carbon power system requires rapid changes in the characteristics and configuration of the electricity system to increase the share of RES generation and reduce the amount of generation from fossil fuels (Thema et al., 2013, p. 15). This implies an expansion of the European transmission grid across borders and the need for back-up capacity and flexibility of the system to counterbalance any potential phase-out of nuclear, the increased intermittency of RES and weather dependency (Thelma et al., 2013, p. 15). Further development of such technological solutions requires policy backing.

Another technological concern relates to capacity, including whether the electricity markets can deliver enough capacity to match the electricity demands (Thema et al., 2013, p. 13). The absence of EU regulation on capacity markets (as of 2014) makes it even more difficult for member states to coordinate their divergent capacity mechanisms. The purpose of capacity mechanisms is to increase capacity and/or flexibility of the system by creating more incentives for investments in generation capacity, and to promote demand-side flexibility. Since capacity mechanisms have been geographically limited to national markets, the Commission has expressed concern that this may affect generation and investment decisions within the EU internal energy market, which in turn could constrain investments in interconnector capacity (Thema et al., 2013,

p. 13). In a 2012 consultation on generation adequacy and capacity mechanisms, the lacking interconnection capacity, the different RES-support schemes and the lacking demand-side response were indicated as the major barriers to effective market functioning (European Commission, 2012b).

Fifth, the socio-economic challenges are associated with the different economic and social conditions in member states that could have an impact on the affordability of new RE technologies and grid extension in the short and medium term. Thus, the vulnerability of some Eastern European countries highly dependent on gas and nuclear could be particularly exacerbated by the financial crisis in Europe and the shutdown of nuclear power plants in some member states after Fukushima.

Socio-economic challenges in the energy transition process are also linked to public acceptance. Since decarbonization measures would also affect the energy market prices, social policy measures are required to protect vulnerable consumers. Furthermore, the long-term implications of decarbonization need to be clearly explained and communicated to European citizens. The public protests in Bulgaria and Estonia in 2013 against the rising costs of electricity bills and the power of energy monopolies in these countries have demonstrated that any significant rise of energy prices in the future could face public opposition (Tsolova, 2013; Hõbemägi, 2013). The Commission's report on EU energy markets in 2011 backs up consumers' fears and indicates an overall increase in retail prices in Europe, with significant differences among member states (European Commission, 2012c). As a result, the European Consumer Organization (BEUC) expressed its concern that consumers face 'sky-rocketing energy prices', problems when switching providers, monopolistic markets and aggressive marketing practices (BEUC, 2012).

Thus, the integration of climate and energy market policy at EU level is not matched by sufficient integration of RE policies and social policy measures to address the regulatory, technological and social challenges of the energy transition process.

3. Explaining the insufficient integration of decarbonization objectives

To explain the insufficient consideration of decarbonization in the EU's internal energy market policy development, I discuss each of the four variables outlined in Chapter 1: functional overlap, political commitment, societal backing and institutional set-up.

3.1 Functional overlap

In the case of the third liberalization package on energy, the mutually reinforcing objectives of internal energy market development and climate policy account for a clear but indirect functional overlap between both policy fields. Increasing the competitiveness of the energy market is expected to optimize the integration of renewables in the power grid, which will contribute to EU decarbonization goals (Dupont & Primova, 2011). However, the specific connections between the development of a competitive energy market and climate policy objectives were not explicitly taken into consideration during the negotiations among the EU institutions.

Furthermore, a well-integrated and functioning market has the potential to support the system change necessary for the energy transition more efficiently and cost-effectively (European Commission, 2012a, p. 12). A supranational regulatory framework could also improve energy market operation and facilitate cross-border trade of electricity. To achieve this, the development of network codes will be of great importance for the operation of truly cross-border and wholesale markets. Network codes could be defined as a set of internal market electricity rules developed by ENTSO-E in cooperation with ACER. The aim is to accelerate the harmonization and effectiveness of the European electricity market.[2] Hence, they will play a significant role in completing the infrastructure to secure energy supply and deliver EU climate change objectives.

The functional overlap between the internal energy market and decarbonization can be considered a significant factor explaining the reinforcement of decarbonization objectives in internal energy market legislation. Although climate policy concerns were not sufficiently acknowledged in the policymaking process, the EU competitiveness target and the improvement of the regulatory framework for electricity trade resulted in synergies between decarbonization objectives and the common market goal. These synergies could be better and more explicitly pursued in energy market policymaking, for the benefit of decarbonization.

3.2 Political commitment

The level of political commitment in internal energy market policy to decarbonization seems very low (Dupont & Primova, 2011). Except for outlining general objectives of the third legislative package as achieving 'a more secure, competitive and sustainable supply' (Council of the European Union, 2008a, p. 6; 2008b, p. 17) and promoting 'sustainability by stimulating energy efficiency and guaranteeing that small companies, too, in particular those investing in renewable energy, will have access to the energy market' (Council of the European Union, 2009, p. 19), no other relevant reference to climate policy objectives could be traced in the statements relating to the internal energy market by the Council of Energy Ministers.

In the Commission's inter-service coordination, environmental concerns played only a marginal role (Primova, 2013a, p. 92). Both DG Energy and DG Competition had active roles in the inter-service coordination and co-drafting of the legislative proposals. This active involvement shaped the final policy proposals (European Commission, 2007a). The issues related to the non-discriminatory access to the grid were not fully discussed during the inter-service coordination of the third energy package but were extensively deliberated in the framework of the proposal for the 2009 Renewable Energy Directive (Directive 2009/28/EC), which was negotiated in parallel. During the parallel drafting of the climate and energy package it was acknowledged that the EU could meet the 20-20-20 targets, if the grid were able to absorb all the renewable energy production.

Furthermore, the importance of the third energy package for guaranteeing access for RE producers and improving energy efficiency was highlighted by some Members of the European Parliament (MEPs) (Dupont & Primova, 2011). Therefore, some limited level of political commitment could be seen in the European Parliament, in particular in the statements of Green MEPs who highlighted the importance of fair market conditions, transparent prices and decentralizing measures to ensure access for RE suppliers. The European Greens showed less satisfaction with the outcome of the Council negotiations than other party groups, deeming the agreed provisions insufficient for dismantling the dominant positions of the energy monopolies, for preventing price rises and for ensuring transparency and consumer protection (Primova, 2013a, pp. 143, 158). The debate on the liberalization of the EU energy sector focused on the issue of ownership unbundling as well as regulatory aspects of the EU internal energy market, but no priority was assigned to climate policy goals.

There is, therefore, only limited evidence of political commitment to the EU decarbonization objectives in the EU internal energy market area during the policy process leading to the third package. The low acknowledgement of relevant climate policy links partially explains the insufficient integration of decarbonization objectives in the energy market sector (that is, the lack of recognition of functional overlap).

3.3 Societal backing: Stakeholder involvement and support

A broad range of stakeholders participated in the policy process leading to the third energy package. However, environmental non-governmental organizations were present only to a limited extent compared to private industry actors (Primova, 2011). The marginal involvement of environmental organizations in the policy preparation of the package and the limited recognition of the link between EU climate policy objectives and an integrated and competitive energy market enabling non-discriminatory access for independent producers of RE led also to less deliberation on this matter (Primova, 2013a). These concerns were, however, addressed during the deliberations in the European Parliament.

The more powerful energy industry lobbies and business associations representing industrial consumers may have had a disproportionate influence on the policy process. They mobilized their resources and used various channels for policy influence – through their national governments, contacts with MEPs and Commission officials. The divergence in stakeholder participation and support corresponds to resource asymmetries: different stakeholder groups have different levels of capacity to monitor the complex EU decision-making process. The Commission attempted a more active approach to include a broader group of stakeholders in the dialogue beyond the traditional players that had previously dominated the energy policy discussions in Brussels. But many environmental groups were deeply involved in the negotiations on the climate and energy package and could not dedicate resources to the third energy package (Primova, 2013a, pp. 382–383).

The marginal involvement of environmental civil society actors at both the policy preparation and the decision-making stage can also be linked to the limited acknowledgement of the link between decarbonization and the internal energy market during the policy process. This energy market itself is an issue that does not garner much citizen interest.

3.4 The institutional and policy context

The institutional and policy context played a significant role in the internal energy market development and the evolution of its decarbonization dimension. The growing importance of energy security issues and the challenge of climate change in the 2000s undoubtedly provided a major impetus for the Europeanization of EU energy policy (Geden, 2008). The political initiative behind the development of a coherent EU energy policy came in the context of rising oil and gas prices in Europe; the increasing dependency of the EU on a few external suppliers; and global climate change challenges – the repercussions of which were discussed during the informal Hampton Court Summit of heads of state and government in the EU in 2005 (Blair, 2005). The summit, organized by the UK presidency at the time, aimed to discuss the role of the EU in a globalized world, with improving EU energy cooperation as one important item on the agenda.

In terms of institutional context, I see the development of the internal energy market policy as a path-dependent process (Dupont & Primova, 2011). The third liberalization package aimed at removing regulatory gaps from the previous two packages and establishing a supranational regulatory framework for cross-border coordination of transmission system operators and national energy regulators (Dupont & Primova, 2011). Furthermore, the internal energy market was one of the pilot cases for the active application of EU competition policy – a priority under the Barroso I Commission mandate (Eikeland, 2008, p. 19). Thus, the internal market programme, competition and environmental policy created spillovers in the energy sector over time, which led to advancing the energy market liberalization agenda (Dupont & Primova, 2011). The policy context that developed provided favourable conditions for launching energy policy initiatives in the EU, facilitating a move towards decarbonization.

Overall, the institutional and policy context, in particular driven by security of supply concerns, environmental threats and the active application of EU competition law, played an important role for incorporating decarbonization elements in the internal energy market agenda. This was due to an enhanced awareness of the need for a coherent energy strategy, putting climate threats on the agenda of EU energy policy discussions, thus aiming to create better conditions for RE producers to compete with energy incumbents. Although decarbonization objectives were not at the forefront of the EU internal energy market agenda, the institutional spillover and the policy context pushed forward some of the decarbonization targets in the internal energy market.

4. Solutions and policy recommendations

In order to tackle the major challenges outlined above, I suggest five main solutions and policy recommendations for future policy on the internal energy market:

1) Enhance coordination across policy fields and among different institutional actors;

2) Improve government support for RE promotion and regulatory certainty;
3) Improve infrastructure development;
4) Encourage technological development and deployment; and
5) Enhance the role of local and regional energy actors in the energy transition process.

First, due to the cross-pillar and cross-sectional nature of this policy domain (Primova, 2013a, p. 379) and its increasing institutional and technical complexity, a more coordinated approach at EU level is needed. Climate and energy policy measures could, for example, be better incorporated in the EU budget review, EU cohesion and regional policies, technology and innovation plans. A European Parliament study on infrastructure for RES concludes that the horizontal cooperation of policy areas within energy supply and distribution needs to be strengthened at the EU level through more intense collaboration of DGs Regional Development, Agriculture, Enterprise, Energy and Climate (Schuh et al., 2012, p. 99). This could enhance complementarity among different policies to promote renewable energy infrastructure.

Second, support for RES and regulatory stability should be part of the strategy for accelerating the uptake of RES in the internal electricity market. The stability of governments' and regulators' commitments, including to support schemes and incentives, are essential to overcome policy uncertainties and to stimulate investments (Capgemini, 2011, p. 9). Furthermore, building the demand side is seen as a crucial element in the decarbonization strategy from an energy market perspective (ECF, 2013, pp. 20–27; Cooperatives Europe, 2013). The demand-side response could be stimulated by allowing new, innovative actors to participate in the energy market, including a wide range of entities from traditional suppliers to electric vehicles manufacturers, exploiting different 'end-use' energy storage opportunities and strengthening the role of national regulatory authorities in driving demand response (ECF, 2013, pp. 20–27).

Third, a pan-European energy policy could also focus on the connection between the trans-European energy grid to small-scale decentralized grids (Schuh et al., 2012, p. 22; see also Chapters 4 and 6). Decentralization is seen as a way to avoid fragmentation of markets and lower the threshold for investments (Cooperatives Europe, 2013). Investments in electricity storage from renewables and infrastructure construction for the electrification of the transport sector are encouraged as further solutions (Schuh et al., 2012, p. 30; see also Chapters 3 and 6).

Fourth, in terms of technological solutions, the promotion of the physical integration of markets and the optimization of system operations are further essential instruments (ECF, 2013, pp. 20–27). Increasing support for research and development in infrastructure, storage and smart grids would be an important way forward for the internal energy market and decarbonization.

Fifth, enhancing cross-border regional cooperation and the role of local and regional actors in the policy implementation process could unlock further opportunities for stimulating RE development, regional electricity market integration, the optimization of resources and improved conditions on the retail markets. Since regional and local actors are expected to be important players in the energy transition process because of the increase in decentralized power generation, the role of energy cooperatives could play a key role in the transformation process and bring more bottom-up citizens' projects to realization. This is something that has not been addressed in-depth in the Commission's energy strategies (Cooperatives Europe, 2013). Furthermore, citizens' co-decision and ownership in the energy transition process not only has the potential to increase public acceptance but also to make the internal energy market more stable, regulate demand and stimulate indigenous supply sources (Cooperatives Europe, 2013).

An important stepping-stone in the consolidation of local and regional RE initiatives is REScoop 20-20-20,[3] founded in April 2012. The federation of existing REScoops (Renewable Energy Sources Cooperatives) in Europe aims to promote local citizen involvement in RES and introduce the REScoop approach to policymakers. Improved regional governance and cross-border regional cooperation can help advance market integration and decarbonization objectives. An ECF study suggests that the formalization of a regional governance structure could optimize cross-border resource sharing and also capture the diverse national contexts and priorities (ECF, 2013, pp. 33–34). Heightened consultation between domestic stakeholders and energy regulatory authorities may bring more of civil society's concerns to the EU arena of policymaking (Primova, 2013a; 2013b). This could broaden the spectrum of local, regional and national institutional and civil society actors in the implementation of energy transition measures.

Any plans for a fourth energy legislative package should therefore focus on specific policy measures for better coordination between the internal energy market goals and decarbonization objectives. Such measures could, for example, address the improved incorporation of EU decarbonization targets into network development plans and the creation of more financial incentives by ACER and national regulators for transmission system operators to deliver their network plans in a timely manner (ECF, 2013, p. 21). Furthermore, the measures should aim at driving an efficient demand response and achieving higher system flexibility in order to allow the absorption of variable energy sources on a larger scale.

Conclusions

In this chapter, I argued that a fully competitive and functioning energy market is an essential condition for achieving EU decarbonization objectives. The analysis has illustrated the contribution that energy market integration can make to EU decarbonization goals through improving competition, removing barriers in cross-border electricity trade, enhancing regulatory governance, creating more incentives for investments in modern energy infrastructure and technology improvements, and decreasing the costs and risks of energy transition. These findings are also in line with the empirical results of the ECF's cost and risk analysis about power sector decarbonization, which shows that an integrated pan-European market is a crucial element in delivering decarbonization objectives (ECF, 2013).

The link between the EU common energy market and decarbonization, being primarily indirect and less evident, has nevertheless been insufficiently addressed throughout the European energy market integration process. It is mostly since the adoption of the third liberalization package on energy and the energy infrastructure package that the links started to become more explicit. Overall, the integration of decarbonization objectives into the internal energy market policy has been low. It has been driven mostly by the functional overlap between different policy objectives and the policy and institutional context. The low political commitment and limited stakeholder involvement helps to explain the low degree of integration of decarbonization objectives in this area. Another reason is the lack of coordinated and coherent policies and measures in the EU energy policy field. The analysis shows that the different dimensions of EU energy policy (environmental, security of supply, market and social considerations) have not been sufficiently recognized in the process of energy market integration. I suggest that strong efforts for a more integrated and coordinated approach are required, so that energy and decarbonization objectives are adequately addressed horizontally in other policy fields but also vertically at different governance levels.

The major drivers for integrating decarbonization objectives into the EU energy market policy have been EU competition rules, EU regulatory governance in the energy sector, the spillover of energy market initiatives, and the functional overlap between the EU competitiveness and sustainability goals in the energy policy field. Taking into consideration the competence, regulatory, infrastructural, technological and socio-economic challenges to internal energy market development, any potential fourth energy package should focus on energy infrastructure development, which should be better aligned with the EU decarbonization objectives.

Enhancing regional governance in the energy sector could offer more flexibility, facilitate market policy integration and the development of network rules, as well as increase public acceptance for energy transition goals and measures. This could be done through the formalization of regional governance structures or by enhancing participatory mechanisms for stakeholders and individual citizens at local and regional level, so that they can become active drivers in the energy transition process. Since the link between the trans-European energy grids and small-scale decentralized grids is seen as an essential element of a supranational energy policy, strengthening the regional level of energy governance could serve as a key bridge in the EU energy strategy.

Notes

1. Interview with the president of EREF, Brussels, 2 March 2012.
2. See http://networkcodes.entsoe.eu/, date accessed 5 May 2014.
3. See http://rescoop.eu, date accessed 20 January 2014.

References

Aalto, P. & Westphal, K. (2008) 'Introduction', in P. Aalto (ed.), *The EU-Russian Energy Dialogue: Europe's Future Energy Security* (Aldershot: Ashgate), pp. 1–21.
Andersen, S. (2000) 'EU Energy Policy: Interest Interaction and Supranational Authority', *ARENA Working Paper 2000*(5), http://www.sv.uio.no/arena/english/research/publications/arena-publications/workingpapers/working-papers2000/wp00_5.htm, date accessed 21 April 2013.
BEUC (2012) 'Press Release: EU Energy Market. Put the Spot on Consumers', 15 November 2012, PR 2012/031.
Blair, T. (2005) 'Opening Statement: Press Conference at EU Informal Summit, Hampton Court', http://webarchive.nationalarchives.gov.uk/+/http://www.number10.gov.uk/Page8393, date accessed 5 May 2014.
Buchan, D. (2010) 'Energy Policy: Sharp Challenges and Rising Ambitions', in H. Wallace, M. A. Pollack & A. R. Young (eds.), *Policy-Making in the European Union* (Oxford: Oxford University Press), pp. 357–380.

Cabau, E. (2010) 'Unbundling of Transmission System Operators', in E. Cabau, Doherty, F. Ermacora, F. Graeper, C. Jones, C. Schoser, C. Silla & W. Webster (eds.), *The Internal Energy Market: The Third Liberalisation Package, EU Energy Law*, Vol. 1, 3rd edn. (Leuven: Clays & Casteels), pp. 87–182.

Capgemini (2011) *European Energy Markets Observatory: 2010 and Winter 2010/2011 Data Set*, 13th edn. (Paris: Capgemini).

Capgemini (2012) *European Energy Markets Observatory: 2011 and Winter 2011/2012 Data Set*, 14th edn. (Paris: Capgemini).

Cooperatives Europe (2013) 'Contribution to the EU Consultation on Green Paper: A 2030 Framework for Climate and Energy Policies', 28 June 2013, https://coopseurope.coop/sites/default/files/Green%20Paper%20%272030% 20Energy%20%26%20Climate%20Policy%20Framework-Responses%20to% 20Questions.pdf, date accessed 14 May 2014.

Council of the European Union (2008a) 'Press Release: 2875th Council Meeting Transport, Telecommunications and Energy', Luxembourg, 6 June 2008, PRES/08/16.

Council of the European Union (2008b) 'Press Release: 2895th Council Meeting Transport, Telecommunications and Energy', Luxembourg, 9 and 10 October 2008, PRES/08/276.

Council of the European Union (2009) 'Press Release: 2953rd Council Meeting Environment', Luxembourg, 25 June 2009, PRES/09/190.

Dupont, C. & Oberthür, S. (2012) 'Insufficient Climate Policy Integration in EU Energy Policy: The Importance of the Long-term Perspective', *Journal of Contemporary European Research*, 8(2), 228–247.

Dupont, C. & Primova, R. (2011) 'Combating Complexity: The Integration of EU Climate and Energy Policies', *European Integration Online Papers (EIoP)*, 15(Special Mini-Issue 1), article 8.

ECF (2011) 'Power Perspectives 2030: On the Road to a Decarbonised Power Sector', European Climate Foundation, November 2011, http://www.roadmap2050.eu/attachments/files/PowerPerspectives2030_FullReport.pdf, date accessed 10 May 2014.

ECF (2013) 'From Roadmaps to Reality: A Framework for Power Sector Decarbonisation in Europe', European Climate Foundation, December 2013, http://www.roadmap2050.eu/attachments/files/Fromroadmapstoreality(web).pdf, date accessed 10 May 2014.

Eikeland, P. O. (2004) 'The Long and Winding Road to the Internal Energy Market – Consistencies and Inconsistencies in EU Policy', *FNI Report* 8/2004 (Lysaker: Fridtjof Nansen Institute).

Eikeland, P. O. (2008) 'EU Internal Energy Market Policy: New Dynamics in the Brussels Policy Game?' *FNI Report* 14/2008 (Lysaker: Fridtjof Nansen Institute).

Eikeland, P. O. (2011) 'EU Internal Energy Market Policy: Achievements and Hurdles', in V. L. Birchfield & J. S. Duffield (eds.), *Toward a Common European Union Energy Policy: Problems, Progress, and Prospects* (New York: Palgrave Macmillan), pp. 13–40.

European Commission (2007a) 'Impact Assessment Accompanying the Legislative Package on the Internal Market for Electricity and Gas', SEC(2007) 1179.

European Commission (2007b) 'Prospects for the Internal Gas and Electricity Market', COM(2006) 841.

European Commission (2007c) 'Inquiry Pursuant to Article 17 of Regulation (EC) No 1/2003 into the European Gas and Electricity Sectors (Final Report)', SEC(2006) 1724, COM(2006) 851.

European Commission (2007d) 'An Energy Policy for Europe', COM(2007) 1.

European Commission (2011a) 'Proposal for a Regulation on Guidelines for Trans-European Energy Infrastructure and Repealing Decision No. 1364/2006/EC', COM(2011) 658.

European Commission (2011b) 'Energy Roadmap 2050', COM(2011) 0885.

European Commission (2011c) '2009–2010 Report on Progress in Creating the Internal Gas and Electricity Market', 9 June 2011, http://ec.europa.eu/energy/gas_electricity/legislation/doc/20100609_internal_market_report_2009_2010.pdf, date accessed 8 November 2014.

European Commission (2012a) 'Making the Internal Energy Market Work', COM(2012) 663.

European Commission (2012b) 'Consultation Paper on Generation Adequacy, Capacity Mechanisms and the Internal Market in Electricity', 15 November 2012, http://ec.europa.eu/energy/gas_electricity/consultations/doc/20130207_generation_adequacy_consultation_document.pdf, date accessed 8 November 2014.

European Commission (2012c) 'Energy Markets in the European Union in 2011' COM(2012) 663.

Geden, O. (2008) 'Die Energie- und Klimapolitik der EU – zwischen Implementierung und strategischer Neuorientierung', *Integration, 4*, 353–364.

Höbemägi, T. (2013) 'Centre Party to Protest against High Electricity Prices in Tartu', *Baltic Business News*, 20 February 2013, http://www.balticbusinessnews.com/article/2013/2/20/centre-party-to-protest-against-high-electricity-prices-in-tartu, date accessed 15 November 2013.

Isabel, M. & Soares, R. T. (2004) 'Restructuring of the European Power Industry: Market Structure and Price Volatility', http://sessa.eu.com/documents/wp/D23.2_Soares.pdf, date accessed 21 November 2009.

Jamasb, T. & Pollitt, M. (2005) 'Electricity Market Reform in the European Union: Review of Progress toward Liberalization & Integration', *Working Paper 5(3)*, Center for Energy and Environmental Policy Research (CEEPR).

Johnston, A. & Block, G. (2012) *EU Energy Law* (Oxford: Oxford University Press).

Jones, C. (2010) 'Introduction', in E. Cabau, A. Doherty, F. Ermacora, F. Graeper, C. Jones, C. Schoser, C. Silla & W. Webster (eds.) *The Internal Energy Market: The Third Liberalisation Package*, Vol. 1, 3rd edn. (Leuven: Clays & Casteels), paragraphs 1.1–1.33.

Newbery, D., Strbac, G., Pudjianto, D. & Noël, P. (2013) *Benefits of an Integrated European Energy Market* (Amsterdam: Booz & Company; London: LeighFischer).

Nilsson, M. (2011) 'EU Renewable Electricity Policy: Mixed Emotions toward Harmonization', in V. L. Birchfield & J. S. Duffield (eds.), *Toward a Common European Union Energy Policy: Problems, Progress, and Prospects* (New York: Palgrave Macmillan), pp. 113–130.

Primova, R. (2011) 'Enhancing the Democratic Legitimacy of EU Governance? The Impact of Online Public Consultations in Energy Policy-Making', *RECON Online Working Paper 2011(1)*, http://www.reconproject.eu/main.php/RECON_wp_1101.pdf?fileitem=5456456, date accessed 29 March 2013.

Primova, R. (2013a) (PhD Thesis) *Assessing Political and Social Accountability in EU Policymaking: A Case Study on the Multi-Level Coordination of the EU Internal Energy Market Policy* (Brussels: Vrije Universiteit Brussel).

Primova, R. (2013b) 'The Complex Governance of EU Energy Policy', in *ecoTips, Trends in Sustainability*, October–November 2013(5), 38–39.

Reiche, D. & Bechberger, M. (2004) 'Policy Differences in the Promotion of Renewable Energies in the EU Member States', *Energy Policy, 32*, 843–849.

Schuh, B., Dallhammer, E., Damsgaard, N. & Stewart, E. N. (2012) *Infrastructure for Renewable Energies: A Factor of Local and Regional Development* (Brussels: European Parliament).

Thema, E3M-Lab & COWI (2013) 'Capacity Mechanisms in Individual Markets within the IEM', *Document TE-2013-06*, Project ENER/B2/175/2012 (Brussels: European Commission).

Tsolova, T. (2013) 'Bulgarians Protest over Jump in Electricity Bills', *Reuters*, 10 February 2013, http://www.reuters.com/article/2013/02/10/us-bulgaria-protests-electiricity-idUSBRE91904Y20130210, date accessed 15 November 2013.

3

The Power Sector: Pioneer and Workhorse of Decarbonization

Stefan Lechtenböhmer and Sascha Samadi

Introduction

In 2012, fossil fuels made up 49 per cent of total gross electricity generation in the EU-28 (Eurostat, 2014). This electricity generation was responsible for more than one-quarter of Europe's greenhouse gas (GHG) emissions (EEA, 2014).

For the power sector, the 'Roadmap for moving to a competitive low-carbon economy in 2050' published by the European Commission (European Commission, 2011a) envisages an almost complete decarbonization by 2050 with reductions of GHG emissions of 93–99 per cent compared to 1990. Emission reductions in the power sector are expected to be considerably higher than in other sectors, where the potential for cost-effective reductions is more limited (see Chapter 1). The realization of such strong GHG emission reduction in the electricity sector has been further explored by the Commission in a specific Energy Roadmap 2050 (European Commission, 2011b) and has also been a subject of several scenario studies by stakeholders. Many of these studies have stressed that decarbonization in the electricity sector will lead to so-called co-benefits for society beyond climate change mitigation, including

- a significant decrease in the share of fossil fuel energy imports (natural gas, coal and oil) and thus an increase in Europe's energy independence by relying more strongly on domestic (renewable) resources;
- an increase in air quality and thus a reduction in health-related impacts of air pollution;
- the creation of millions of new jobs in Europe due to investments in new technology and infrastructure.

Whereas there are multiple reasons for Europe to decarbonize its electricity sector, this chapter takes a closer look at *how* such a decarbonization can be achieved within a time span of three to four decades. To this end, we compare key electricity sector decarbonization scenarios available in early 2014. We focus on analysing the common strategies they *all* foresee on both the demand and supply side to achieve the deep emission reductions required to 2050. On this basis, we identify key policy measures that need to be in place to implement successfully those demand- and supply-side strategies. Finally, we briefly discuss the main drivers of, and barriers to, decarbonization of the power sector.

1. Scenarios for a decarbonized power sector

Table 3.1 describes six studies with a total of 25 scenarios of the EU's electricity system development until the year 2050. While the aims of the respective studies differ to some extent, all of them focus on how to decarbonize the EU electricity system by 2050. The table provides the names of the studies, the institutions that commissioned the studies and the respective dates of publication. As most of the studies comprise of more than one scenario, the name of the decarbonization scenario selected for our analysis is provided and the core assumptions are listed. Many of the scenarios differ considerably in their assumptions on which energy sources and technologies will be used to generate electricity in a future decarbonized electricity system. Scenarios also differ with regard to the level and sector-by-sector structure of future electricity demand.

However, all scenarios interestingly rely on a few identical key strategies. We identify two such key strategies each on the demand and on the supply side. There is wide agreement in the scenario literature that these four common strategies will need to be implemented in order for the European power sector (as well as the overall energy system) to be decarbonized.

2. Key strategies on the demand side

2.1 Reducing 'traditional' electricity demand

Although not always highlighted, it is a core assumption in all decarbonization scenarios that European societies will use electricity in an increasingly efficient way. In the reference scenarios of all analysed studies, a further increase in 'traditional' electricity demand in the residential, commercial and industrial sectors (such as household

Table 3.1 Scenario studies of the European energy and electricity system to 2050

Name of the study	Reference	Selected decarbonization scenarios(s)	Core assumptions (for selected scenario)
Power Choices Reloaded	Eurelectric (2013)	'Power Choices Reloaded' (one of six scenarios)	• High acceptance in society for nuclear power and CCS • CCS to be technically viable and economically attractive
Energy [r]evolution – A Sustainable EU 27 Energy Outlook	Greenpeace & EREC (2012)	'energy [r]evolution' (one of two scenarios)	• Nuclear power to be phased out in Europe • CCS not used • Far-reaching decarbonization through efficiency and renewables, including renewable electricity imports from North Africa
Energy Technology Perspectives 2012	IEA (2012)	'2DS' (one of three scenarios) (The name of the scenario refers to its CO_2 emissions, which are expected to be in line with limiting global warming to 2°Celsius by 2100 over pre-industrial levels.)	• High acceptance in society for nuclear power and CCS • CCS to be technically viable and economically attractive
Energy Roadmap 2050	European Commission (2011b, c, d)	'Diversified Supply', 'Energy Efficiency' and 'High RES' (three of seven scenarios)	• High acceptance for all technologies (Diversified Supply) • Strong energy efficiency improvements (Energy Efficiency) • Renewables favoured in electricity sector over nuclear power and CCS (High RES)

Tangible ways towards climate protection in the European Union (EU Long-term scenarios 2050)	Fraunhofer ISI (2011)	'A' and 'B' (two of two scenarios) (The two scenarios differ only in the assumed development of electricity demand.)	• Nuclear power to be phased out in Europe • CCS not used • Far-reaching decarbonization through efficiency and renewables
Roadmap 2050 – A Practical Guide to a Prosperous, Low-Carbon Europe	ECF (2010)	'40% RES' and '100% RES' (two of five scenarios)	• 40% RES scenario assumes only 40% share of renewables in electricity generation in 2050 and 30% each for nuclear and CCS • 100% RES assumes all renewables, including 15% imports from North Africa

Note: CCS = Carbon Capture and Storage.
Source: Own table, information taken from the studies mentioned.

appliances, machinery, lighting and cooking) by 37–48 per cent by 2050 is expected (compared to their respective base years). In contrast, in the decarbonization scenarios, this 'traditional' electricity demand is expected to remain relatively stable until 2050 or even *decline* by up to 15 per cent. Hence, by 2050, 'traditional' electricity use needs to be cut by about 30–45 per cent compared to a reference development (see the difference between reference and decarbonization scenarios in Figure 3.1). The implied strong decoupling of economic growth from 'traditional' electricity demand appears to be technologically feasible, according to many decarbonization scenarios, but will require effective policies for achieving a much more efficient use of electricity.

2.2 Substituting electricity for fossil fuels

All scenarios expect low-carbon electricity to expand into new applications, such as road vehicles and efficient low-temperature heat generation (for example, heat pumps in buildings) and certain industry processes (see Figure 3.1). The final energy carrier introduced will thus be increasingly based on low or zero GHG emissions. Electric motors and heat pumps are often more efficient than the fossil

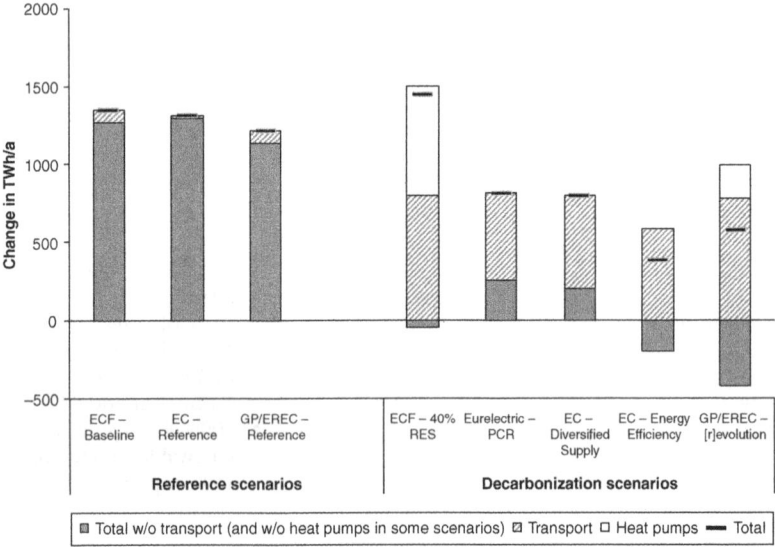

Figure 3.1   Changes in electricity demand for traditional and new appliances, comparison of scenarios for 2050, EU-27

Note: 'total w/o transport (and w/o heat pumps in some scenarios)' is without additional electricity demand from heat pumps in the scenarios 'ECF – 40% RES' and 'GP/EREC – [r]evolution'; ECF = European Climate Foundation, EC = European Commission, GP = Greenpeace, EREC = European Renewable Energy Council.

Final electricity demand EU-27, 2010: 2844 TWh (Eurostat, 2014).

Source: Own figure based on data from the mentioned scenario studies.

fuel-based technologies they substitute. Therefore, the electricity system will not only decarbonize itself, but it is also expected to 'export' decarbonization to other sectors by substituting fossil fuels with electricity.

For the transport sector, the Greenpeace and EREC (2012) study is most ambitious by assuming that by 2050 electricity will supply roughly 80 per cent of the sector's final energy demand, either directly (50 per cent) or indirectly (30 per cent) via hydrogen produced from electrolysis. This scenario requires about 75–80 per cent of all light-duty vehicles in the EU to use electric propulsion by 2050. Furthermore, a large share of the medium and heavy-duty vehicles would run indirectly on electricity, by using hydrogen (see Chapter 6).

In buildings, electricity will also increasingly substitute fossil fuels, for example, through electric heat pumps, which also reduce final overall energy demand (compared to traditional heating technologies) given that they make use of environmental heat.[1] In the European Commission's Energy Efficiency scenario, the share of electricity in energy supply for heating and cooling will increase from less than 10 per cent in 2010 to more than 20 per cent of final energy demand in 2050. If the electricity for heating purposes were to be fully used by heat pumps, these would supply more than 40 per cent of the heat demand of all buildings (see also Chapter 7).

In addition to road vehicles and heating, many industrial processes could be converted from fossil fuels to electricity. While the majority of fossil fuels in the industrial sector is used for high-temperature processes, which are not suited to heat pumps, in some industrial processes fossil fuels can be substituted by electricity through the use of other technologies such as infrared heaters. Other processes in the chemical industry and in steel making could be converted to the direct or indirect use (via hydrogen) of electricity. These processes are, however, in many cases not more energy efficient than conventional processes. This may be why the scenarios discussed here do not assume a significant switch from fossil fuels to electricity in the industrial sector (see Chapter 5).

3. Key strategies on the supply side

3.1 Increasing low-carbon electricity generation

Electricity sector GHG emissions in the analysed scenarios are reduced by at least 90 per cent by 2050 compared to 1990. This requires that electricity generation from non-CCS fossil fuel plants, which in 2012 accounted for nearly half of electricity generation in the EU-28, is either entirely or mostly phased out by 2050. Such a drastic change in Europe's electricity supply structure requires adequate framework conditions for new investments, especially given the long lifetime of power plants and infrastructure like electricity grids (see Chapter 4).

In general, there are three different options for non- or low-carbon electricity generation: nuclear power plants; fossil fuel power plants equipped with CCS technology; and renewable energy sources (RES). In 2012, nuclear power plants generated 27 per cent of the electricity produced in the EU-28 (Eurostat, 2014), but social acceptance of this technology in many European countries is comparatively low (Eurobarometer, 2010; OECD, 2010). Investment costs of newly built nuclear power plants have increased over the years (see Grubler, 2010). Regarding CCS, there are currently no commercial power plants equipped with this technology in the EU and most of the CCS demonstration projects originally planned have been abandoned or put on hold due to difficulties in applying CCS technology on a large scale, costs or questions regarding the social acceptance of CO_2 piplines and storage sites (von Hirschhausen et al., 2012).

RES, however, have significantly increased their contribution to electricity generation in the EU-28 from less than 15 per cent in 2000 to 24 per cent in 2012 (Eurostat, 2014). In all analysed decarbonization scenarios, electricity generated from RES will become the most important low-CO_2 supply option by 2050 with shares of typically 60 per cent or more in 2050. Understandably, the share of renewables in 2050 is higher in scenarios which rule out or limit the use of CCS and nuclear power. However, even in scenarios with more optimistic assumptions regarding CCS and nuclear power and assuming a phase-out of support for renewables, electricity generation will be mostly based on RES by the middle of the century.

Figure 3.2 shows the development of the share of renewable electricity generation in Europe between 2012 and 2050 in the analysed scenarios.

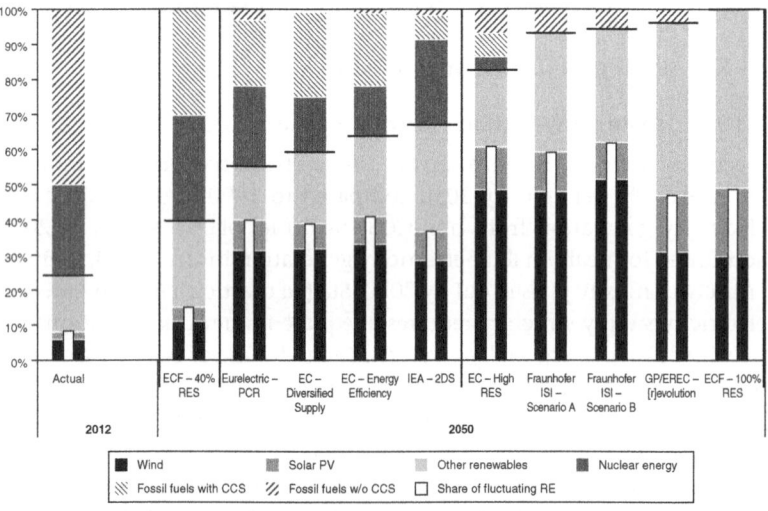

Figure 3.2   Shares of various sources of electricity in meeting European gross electricity demand and shares of fluctuating renewables in 2012 (actual) and in 2050 according to scenarios

Note: EC = European Commission, GP = Greenpeace, IEA = International Energy Agency; the three 'worlds' mentioned in the text and separated based on the share of all renewables in total electricity generation are indicated by the vertical lines separating the decarbonization scenarios.

Source: Own figure based on data in the mentioned scenario studies and on Eurostat (2014) for actual 2012 data.

Looking at the shares in the year 2050 suggests that three diverging 'worlds' of the future European power supply can be differentiated – with specific challenges for each world.

A future with renewable shares of 80 to 100 per cent by 2050 will have to deal most strongly with fluctuations in electricity generation from wind and solar photovoltaic (PV) (see below). While this challenge is likely to be somewhat smaller in the 55–65 per cent renewables world, this world faces the additional challenges of ensuring economic viability and social acceptance for CCS power plants, infrastructure and storage sites, and new nuclear power plants. Finally the single scenario 'outlier' with a more moderate renewables share of 40 per cent will likely face few problems with the fluctuating nature of renewables, but even greater challenges related to the massive deployment of CCS and nuclear power.

Finally, we question whether scenarios with a renewables share of less than 80 per cent in 2050 actually describe a *sustainable* energy system. There are safety, waste and proliferation risks associated with nuclear power (Schneider & Froggat, 2013; Adamantiades & Kessides, 2009). CCS raises sustainability concerns due to uncertainty about the long-term viability of CO_2 storage in the ground and because resource depletion and air pollution remain unaddressed (Viebahn et al., 2012).

3.2   Integrating very high shares of fluctuating electricity generation

Key to transforming the European electricity system by 2050 is the ability to handle a strongly increased share of fluctuating power generation from intermittent RES, most notably wind (onshore and offshore) and solar PV. In 2012, the combined share of wind and solar PV was about eight per cent (Eurostat, 2014), thus more than 90 per cent of electricity came from 'dispatchable' plants that either generate electricity continuously or that can more or less flexibly increase or decrease their output to adapt to changes in electricity demand or power generation in other plants (Eurelectric, 2011).

In all but one of the analysed scenarios, the share of fluctuating electricity generation from wind and solar PV will increase to about 40–60 per cent by the middle of the century (see Figure 3.2). This huge increase will require significant changes in the provision and consumption of electricity if a stable electricity supply is to be maintained. The key challenge is ensuring that electricity demand and supply match at all times.

To this end three main options are available in the short to medium term to integrate growing shares of fluctuating renewables into the electricity system:

- First, it could be necessary to maintain a sufficient stock of flexible, dispatchable power plants that can be easily started and shut down when required (Eurelectric, 2011). Some technologies for this purpose include pumped hydro storage plants and fossil fuel power plants (with natural gas open cycle power plants being the most flexible). While potential sites for hydro storage plants are limited, the use of fossil fuels in dispatchable power plants in the long run contradicts the decarbonization target. Alternatives like biogas, CCS equipped or nuclear power plants either face limited potential or – in the case of CCS and nuclear power plants – high investment costs, which would make electricity generation extremely expensive if these plants are used only for a few hours a year.
- Second, demand can be made more flexible as electricity demand in many sectors can be shifted back or forth a few hours. This is often referred to as *demand-side management*. For example, laundry machines could be run (manually or automatically) during times of the day when electricity generation is high. Household and especially commercial freezers could cool down by a few degrees more than necessary during times of high availability of power, in order to use less or no electricity when electricity supply is temporarily short (Department of Energy & Climate Change, 2012). Real-time changes in electricity prices and an expansion of smart IT technologies could incentivise such load shifting. Some of the scenarios assume that demand-side management is used to a certain extent.
- The third main option is to extend and improve the existing high voltage electricity transmission grid. This option has significant potential and is relatively inexpensive. It is therefore a preferred option in all decarbonization scenarios. The advantage of building a European 'super grid' would be that the broader spatial scope of the electricity system will smooth out fluctuations in both electricity generation and demand. For example, cloudy conditions or a wind lull in some part of Europe could be balanced by sunny skies and strong winds in other parts of Europe. Furthermore, the available flexibility options, like relatively cheap storage in pumped hydro power plants, can be used more effectively if temporary electricity surplusses can be transferred easily from one European region to another (see Chapter 4). Many studies suggest expanding the power grid between

Norway and Central Europe to transfer excess wind or solar power generation at certain times of the day to Norway, where it could be stored in pumped hydro storage power plants (see also Chapter 11).

Summing up, in the short to medium term the most promising option to deal with increasing fluctuating electricity supply is to extend the capacity of the European electricity transmission grid. This would balance fluctuations over longer distances and better use storage and flexibilty options. However, the potential of this and the other options mentioned above are limited. At some point in time these options alone will no longer be sufficient to integrate high shares of fluctuating electricity supply into the system (Denholm & Hand, 2011). Consequently, it may be necessary in the longer term (probably beyond 2030–2040) to invest in storage capacities like batteries, pumped hydro storage plants or – for large-scale and seasonal electricity storage – electrolysis plants. Electrolysis plants transform electricity to hydrogen during times of excess electricity. Hydrogen can be stored relatively easily and can be transformed back into electricity in fuel cells or in combustion plants. Alternatively, it could be used as a low-carbon fuel in transport or industrial processes. However, transforming electricity to hydrogen for storage purposes is expensive and leads to significant energy losses of around 30–35 per cent during the transformation and additional losses of at least 20–30 per cent when the hydrogen is used for electricity generation (Joint Research Centre, 2011).

4. Policy framework and required reforms

The electricity sector is assigned a major role in the decarbonization of the European economy. It is therefore important to take account of the current institutional and policy framework of the sector to assess the ability to initiate the targeted developments. In this section, we discuss the existing energy and climate policies and instruments applied in Europe, following the four core strategies identified above. However, the reference scenarios of all studies analysed make it clear that the current policy framework is insufficient to induce the transformational changes required for the electricity sector to contribute adequately to Europe's decarbonization.

4.1 Strategies to curb growth of traditional electricity demand

Using electricity in an increasingly efficient way is the first strategy for decarbonization. This field has been intensively covered by EU legislation. The Energy Services Directive (2006/32/EC) from 2006 set a framework for energy efficiency, which was subsequently strengthened by the 2012 Energy Efficiency Directive (2012/27/EU). These two Directives oblige member states to set up National Energy Efficiency Action Plans, to set indicative energy savings targets, and to implement specific policy instruments, including obligations for energy supply companies to achieve 1.5 per cent of additional energy savings each year.

The energy efficiency of specific equipment and appliances is regulated in the EU by the Ecodesign Directive (2009/125/EC), which imposes minimum efficiency standards, and the Energy Labelling Directive (2010/30/EU), which mandates that energy efficiency labels are used for a range of products. The coverage of both Directives is increasingly expanded to most energy-using and energy-related products and the stringency of their provisions is subject to regular review.

Furthermore the Energy Performance of Buildings Directive (2010/31/EU) requires member states to set minimum energy performance standards for new buildings and major renovations. From 2021 on, all new buildings in the EU shall be nearly-zero-energy buildings. It also requires the creation and use of energy labels for individual buildings, called Energy Performance Certificates (see Chapter 7).

Nevertheless, the reference scenarios prepared regularly for the European Commission (including one released in early 2014), do not expect a reduction in 'traditional' electricitiy use in the residential, commercial and industry sectors, but only a slow-down of demand growth (European Commission, 2014a). This indicates the need to make better use of the existing energy efficiency instruments and to complement them with new instruments.

For electricity saving in traditional stationary demand, an improved policy framework could consist of:

- Targets for electricity demand reductions consistent with overall energy savings targets. For example, Germany has a target of reducing primary energy demand by 50 per cent by 2050 (compared to 2008) and another target to decrease electricity demand by 10 per cent by 2020 and by 25 per cent by 2050 (Bundesregierung, 2010).
- Continuous adaptations to the national building regulations towards the target of a nearly zero-energy building standard, while aiming for cost-optimal renovation of the building stock. Building regulations could also promote solar water heating which could help replace electric water heating (see Chapter 7).

- Strengthened system of labelling and minimum standards, including through more frequent and tougher (possibly automatic) procedures for tightening the requirements for the labels and of the minimum standards. This could follow a 'top-runner' approach to make sure that technical improvements are incentivized and conveyed to the markets for electricity-using products as quickly as possible. Japan has applied such a regime since the late 1990s, so that minimum efficiency standards are increased regularly, based on the performance of the most efficient appliance on the market (Kimura, 2010).

In the industrial sector, energy use is very diverse and difficult to address specifically, which means that the existing energy efficiency instruments have considerable limitations. Companies often give only low priority to incremental improvements (like electricity savings). Efficiency policy instruments covering crosscutting technologies, such as motors, drives, pumps and lighting, should deal with the specific situations of individual companies. Examples include mandatory energy audits linked to electricity tax exemptions or schemes designed to reduce payback times for electricity saving measures, such as targeted subsidies or the support of dedicated third party funded systems. Such instruments, however, need to be implemented nationally, although they could be indirectly supported by the EU, for example via energy savings targets set by the Energy Efficiency Directive.

4.2   Decarbonization through new electricity uses in transport and heating

Using ('green') electricity instead of fossil fuels in transport and heating requires policies and measures that address several different energy carriers. A broad discussion about the most suitable instruments for promoting electric mobility has developed since the late 2000s (see, for example, Kley et al., 2010; Mock & Yang, 2014). Despite various instruments already in use in many member states, it is difficult to predict whether electric mobility can expand as much by 2050 as envisaged in the low-carbon scenarios. Gaps in market deployment strategies are greater for light duty vehicles, and particularly for lorries, for which technologies for electrification (direct or via 'electro fuels' like hydrogen) are far behind those for passenger transport. If technology development significantly reduces battery costs, market development will soon rely on the cost ratio between electricity and fossil fuels. Given the high share of taxes on road transport fuel prices, there is sufficient scope for appropriate policies favouring electric cars, if the reduced income from fuel sales taxes can be handled (see Chapter 6).

Cost differences have a more limited potential in driving the sub-stitution of fossil fuels by electricity in stationary applications. As tax rates on fuels for heating are much lower, further policies and mea-sures are needed to make the use of electricity economical for individual households and businesses. Building codes setting integrated targets for maximum final energy demand or for maximum GHG emissions, for example, could set incentives at least for new buildings to use highly efficient heat pumps.

4.3   Expansion of low-carbon power generation

Many European countries have introduced support mechanisms for electricity generation from RES, like feed-in tariffs (REN21, 2014). Some European countries are using feed-in-premiums, meaning that operators of renewable energy technologies need to sell their electricity on the market while receiving an additional 'premium' on top of the market price. An alternative instrument currently being used in fewer European countries (for example, Sweden and Poland) is quota obligations, where either utilities or retail suppliers need to make sure that by a specific future date a certain share of the electricity they sell is sourced from RES. The obliged companies may either invest in renewable energy plants themselves, or they may buy 'green' electricity from operators of renewable energy plants.

In most countries, feed-in tariffs have proven very effective in promot-ing the deployment of RES (Haas et al., 2011). However, critics point to the lack of incentives for operators to take account of differences in the value of electricity depending on where, and especially when, it is gen-erated (Nagl, 2013). Policy schemes that link an operator's income to the market price of electricity (as in the case of feed-in-premiums or quota obligations) offer incentives to operators to take the value of electric-ity into account in their investment and operating decisions. However, such policies also increase an investor's risk and thus make it more dif-ficult and more expensive to receive financing, which (by itself) would make the power system transformation more expensive and might make it more difficult for smaller investors to participate.

There is no consensus in the literature on the best policy instrument to support the deployment of RES. A mix of instruments addressing the specifics of various renewable energy technologies and adapted to cer-tain phases of the electricity system transformation might be best suited. Despite falling prices, some financial support for renewable energy technologies will still be needed in the short to medium term, as in most cases electricity generation from renewables remains somewhat more expensive. This is not least because fossil fuel and nuclear power generation does not cover all costs that accrue to society (referred to in economic terms as a 'lack of internalization of external costs') (Owen, 2006).

Another reason why continued support (or perhaps rather 'refinancing') of renewable energy technologies may be required is because growing shares of power from wind and solar PV increasingly drive down wholesale electricity prices. This may lead to the 'missing money' problem. Electricity becomes so cheap on average on the market (because the variable generation costs of wind and solar PV are almost zero and so wholesale prices are very much driven down at times of much wind or sunshine) that operators do not earn enough from selling electricity to cover the plants' investments costs. This may be true for RES and also for conventional technologies (Cramton & Ockenfels, 2011). Consequently, in many European countries, the introduction of capacity mechanisms is being considered. Such mechanisms provide incentives for holding available generating capacity (Sweco Energuide AB, 2014).

In summary, effective policies are already in place in many European countries to support the deployment of RES. The challenge to 2050 in these countries will be to maintain the momentum of deployment, while ensuring that policies are as cost-efficient as possible. Issues concerning public support for renewables should be taken into account when adapting existing support policies. At the same time, the policies in those European countries which have been lagging behind in deployment will have to be considerably improved in the coming years, taking advantage of not just the much lower technology prices compared to a decade or two ago, but also of the experiences made with various policies in other European countries.

Nuclear power has seen very little investment in Europe since the late 1980s (Schneider & Froggat, 2013). Nuclear power plants need public support for new investments to come forward. New nuclear power plants being built in Finland and France and being planned in the United Kingdom in 2014 each take advantage either of investment subsidies or of guaranteed feed-in-tariffs. A feed-in-tariff of at least 89.50 pounds/MWh (or about 10.7 Eurocent/kWh at an exchange rate of 1.20 Euro per British Pound) was agreed upon in 2013 by the British government and the consortium that is planning to build a new nuclear plant at Hinkley Point, Southwest England (Department of Energy & Climate Change, 2013). This tariff is higher than feed-in-tariffs paid to onshore wind in most European countries and higher even than solar PV tariffs in some countries (Prognos, 2014). If new nuclear power plants are considered part of the decarbonization equation, then considerable public financial support will be required. However, we consider that the risks, costs and the very limited compatibility with the needs of fluctuating RES exclude new nuclear power plants as an option.

We also have reservations about the sustainability of CCS power plants. However, if this technology is seen to be an important option for decarbonizing the power sector, the first step would be to support research and development and demonstration projects. Planning in the EU for such demonstration projects has been largely unsuccessful as of 2014 (Global CCS Institute, 2014), which makes the use of CCS technology unlikely until 2025 or 2030 at the earliest. Due to the existing technological and economic uncertainties, we argue that CCS cannot be relied upon as a key decarbonization technology.

All decarbonization technologies would benefit strongly from an adequate internalization of external costs of fossil fuel power generation. This is most relevant for fossil fuel GHG emissions. If emitters were to pay for their contribution to climate change, fossil fuel power generation would be considerably more expensive, as CO_2 prices would have to be well in excess of the low prices of less than ten Euro per tonne of CO_2 as in the European Emission Trading Scheme (ETS) in 2012 and 2013. Rather, the CO_2 price would probably need to be around 30 to 50 Euro per tonne, possibly even higher (Interagency Working Group, 2013). A higher CO_2 price would considerably increase the market competitiveness of renewables, nuclear power and CCS technology. To achieve an appropriate price for CO_2 emission allowances within the ETS, European policymakers need to ensure adequately ambitious emission limits.

4.4 Improving system integration of fluctuating renewable energy sources

Finally, to make high shares of fluctuating RES electricity feasible in the EU electricity system it is necessary to improve their system integration.

Although policies to promote the deployment of RES have been in place for several years in many European countries (as of 2014), the need for policymakers to improve the system integration of fluctuating RES only became apparent in the 2010s. Shares of fluctuating RES in electricity generation have increased quickly and, in 2014, are beginning to have tangible effects on the overall electricity system. Policies should aim at making sure that the future electricity system becomes as flexible as possible. Flexibility on both the supply and demand sides helps integrate significant shares of fluctuating renewables at low cost (IEA, 2014).

Demand-side flexibility can be improved by initiating changes of the electricity infrastructure, for example mandating the installation of so-called smart meters. Such meters can measure electricity consumption and are able to send and receive information from the electricity grid, like the price for electricity at any point in time. It should also be mandated that tariffs are offered to consumers that reflect the changes in wholesale prices at different times of the day. Installing smart meters and offering time-dependent electricity prices would set incentives for consumers to use more electricity when it is abundant (for instance, because of strong winds or lots of sunshine) and to use less electricity when it is scarce. Policy efforts to improve demand-side flexibility will thus require the mainstreaming of policies across multiple fields.

On the supply side, policies that favour more flexible power plants (such as new natural gas power plants) over less flexible power plants (like existing coal and nuclear power plants) should be implemented. For physical and technical reasons, electricity generation in new natural gas power plants can be increased and decreased relatively quickly and at relatively low cost – making it a well-suited technology to complement fluctuating electricity generation from renewables until more flexible demand, an improved electricity grid and additional storage capacity enables a fully renewable-based electricity system (Eurelectric, 2011). Biogas and hydro (reservoir) plants similarly have the potential to adapt electricity production quickly. The policy framework should increasingly incentivize building more flexible power plants – or making existing plants more flexible. Specifically, financial support for biogas power plants (for example, by using feed-in-premiums) could ensure that they generate electricity only when electricity prices are high. In addition, biogas plants could be required to maintain gas storage to enable sufficient operating flexibility.

An important and relatively cheap option to increase flexibility is the improvement of the European electricity grid. The EU should take a leading role in coordinating the required cross-country enhancements. While it has taken initial steps in this direction (European Commission, 2014b), many European countries need to enhance grid capacity within their borders. They may learn from each other how best to plan and implement these enhancements.

A successful energy system transformation requires policy mainstreaming across multiple policy fields. For example, ensuring low-carbon electricity supply options are competitive is not sufficient if complicated local planning procedures make investments unattractive. Also, increasing demand-side flexibility will require not only smart grids but also smart meters and – to some extent – smart applications. For these grids, meters and applications to communicate with each other, policymakers need to ensure certain technical standards are in place ahead of their widespread deployment.

Finally, storage technologies may play a crucial role for integrating large shares of fluctuating RES. Today, most storage options are still very expensive, but further research, for example in the field of batteries and long-term storage in the form of hydrogen, are likely to drive down prices in the future. Targeted demonstration and deployment policies for the various types of storage technologies may be needed to realize the potential cost reductions, but the overall costs of any such support instruments need to be assessed carefully. In any case, governmental support for research and development of new low-carbon or carbon-free power generating technologies as well as storage options is crucial (Vasconcelos et al., 2012). As transformational processes require more than merely technological changes (Leach et al., 2012; Seyfang & Haxeltine, 2012), support for research should also be directed towards a better understanding of the economic and social changes required for and brought about by a transformation towards a decarbonized electricity system.

5. Key drivers and barriers for successful decarbonization

The key strategies for transforming the electricity system are quite clear in most studies. In spite of already having most of the general policies and instruments needed for such a transition in place, their respective level of ambition needs to be significantly increased to achieve a successful transformation. Huge investments in new technology need to be made in the decades leading to 2050.

The required system transformation will create enormous opportunities for European societies and stakeholders but will certainly also face significant resistance. In this section, we analyse the key drivers and barriers for successful decarbonization of the electricity system.

First, there are synergistic overlaps between the necessary expansion of the electricity system infrastructures and the policies towards the completion of the internal energy market (see Chapter 2). The internal market policies and the linked programmes to expand connections between national electricity grids and core Pan-European infrastructures are the driving forces for the expansion of the electricity system.

However, there is no clear European vision of a future electricity grid that would be capable of efficiently balancing electricity generation from fluctuating renewables. This indicates that European policymakers have so far not fully exploited the synergistic overlaps between efforts to complete the internal energy market and efforts to improve the integration of fluctuating electricity generation into the European electricity system.

Another field of overlap is the funding of the transformation of the electricity system. Here, the discussion in 2014 is dominated by efforts to keep energy prices low and reduce burdens on final consumers and tax payers in order to maintain and improve (short-term) competitiveness and increase consumers' non-energy purchasing power. Despite a widespread acknowledgement of the long-term economic benefits of electricity system decarbonization, short-term concerns dominate the actual policy discussion. This can be seen in the discussions about Europe's 2030 energy and climate targets where competitiveness concerns feature prominently (see also Chapter 1).

Second, this focus of discussions also indicates that the political priorities have changed since the end of the 2000s, not least due to the economic crisis. Short-term costs dominate the political discussions in 2014 over long-term economic and environmental benefits. Furthermore, the success that has been achieved, particularly in expanding the use of RES in electricity generation, is beginning to affect significantly and negatively the business interests of the incumbent electricity companies, particularly in countries where the transformation has been quick. These influential actors often demand a political rollback from decarbonization objectives (Cardwell, 2013), in spite of their public commitments in favour of decarbonization (Eurelectric, 2009).

Third, the increasing political difficulties to support the transformation process can also be partly explained by changing priorities of the public (Eurobarometer, 2014). This is due to the effects of the economic crisis and also because, in some countries, consumers now bear noticeable costs for decarbonization. Furthermore, there appears to be a split in public opinion in Europe on decarbonization. While many people in Sweden, Denmark and Germany for example, believe climate change to be a very serious problem (Eurobarometer, 2014) and are more likely to back ambitious decarbonization policies, constituencies in some Central and Eastern European member states have other priorities and are thus – motivated by fears of economic burdens – more inclined to back the interests of established energy players.

Widespread public support can lead to a favourable and stable policy environment and thus is crucial for the transformation process. This is important for ensuring that industry buys into the process and that the necessary investments are realized. Setting credible and ambitious medium- and long-term political targets for the transformation of the energy system is an indication of the existence of widespread public support.

A full discussion of how public support can be fostered is beyond the scope of this chapter. One key element is, however, communication. Policymakers and environmental organizations need to inform the public that a decarbonized electricity system is crucial to mitigate the effects of climate change, but that it also has a number of 'co-benefits', like additional employment opportunities or reductions in air pollution, energy import dependence and nuclear risks (European Commission, 2014d).

Fourth, from an institutional perspective, the EU only has limited competence over the power sector. The EU also has no competence in fiscal matters so that introducing taxes (for example, on conventional energy sources) would require unanimous support by all member states. Regarding the demand side, a comprehensive set of regulations exists at the EU-level, but there is a lack of ambition and sufficient funding mechanisms for investment. For the supply side, the EU has only indirect competencies: through the ETS, internal market policies and other economic policies. The vision for a decarbonized power supply, however, is strictly limited by the right of all member states to choose their preferred energy mix. This is why the EU's Energy Roadmap (European Commission, 2011a) is 'technology' neutral and develops a number of scenarios that vary considerably in the respective shares of low-carbon power generation technologies. It could be crucial that the EU gain more competence in implementing certain strategies to achieve decarbonization – regardless of national energy political priorities.

Conclusions

The Low-Carbon Economy and Energy Roadmaps by the Commission as well as several scenario studies commissioned by various stakeholders all assign a major role to the power sector for achieving the decarbonization of Europe's economy. They expect the sector to speed up its development towards low or zero carbon electricity and also to deliver low-carbon electricity for energy services such as transport and heat generation.

The fact that the power sector is at the centre of the transition to a European low-carbon economy highlights the importance of adequate political regulation and of market actors' willingness and capability to adapt. The analysed scenario studies have shown that four core strategies need to be implemented simultaneously with the help of suitable policy measures. These four strategies need to be integrated into a target-oriented and consistent long-term pathway for the transformation of the overall energy system.

To achieve this transformation, several energy policy fields need to function together, with competencies on both the EU and member state levels. These policy fields include boosting energy efficiency, introducing new electricity-powered applications (like electric cars), expanding renewable electricity generation and improving the integration of fluctuating RES into the European electricity system. To enable an efficient and economically feasible transition, policies to advance an integrated European energy market as well as to improve energy security should be aligned with the decarbonization strategy.

While the expansion of energy infrastructure and a stronger coupling of national electricity systems serve both the completion of the internal energy market and the integration of higher amounts of fluctuating generation, further development of liberalized and unbundled electricity markets may have to be adapted to cope with the new challenges that a system with higher shares of renewable generation poses (Lechtenböhmer & Luhmann, 2013).

The power sector strategies discussed here also significantly contribute to increasing security of supply over the long term, by reducing energy demand and by increasing the shares of domestic energy production. But the power sector transformation requires significant efforts and resources, which also means that vulnerabilities may increase during certain phases (for instance, loss of economic power of traditional energy companies). As security concerns are often short-term and supply-oriented, or are targeted at protecting strategic domestic companies, trade-offs in policy targets may emerge in the short-term and need to be closely monitored.

Finally, the main challenge, as we see it, for the decarbonization of the power sector is the high level of ambition and integration needed for the long-term transformation. Several EU level policies need to be brought in line, which should in principle be possible as they overlap synergistically, but the implementation needs strong backing and political will from all member states. As of 2014, however, the member states have not yet decided on a joint vision to decarbonize Europe's power sector. Some governments are favouring RES, while others emphasize the role of nuclear power and CCS-equipped fossil generation. These differences impede agreement on a shared long-term vision for Europe's electricity supply. Many governments also appear to be more concerned about the short- to medium-term costs of the transition than convinced of its potential benefits. These concerns partly explain why member states frequently do not pursue agreed policy targets, such as improved demand-side efficiency, with sufficient intensity.

Therefore, it remains a major challenge for the EU to increase momentum in the power sector transition towards a much more renewable, flexible and efficient system. A clear vision that could build upon the existing scenarios needs to drive this momentum. Such a vision could help convince the sceptical part of the public and the (incumbent) energy industry of the long-term benefits of the transition. It could determine the long-term policy direction needed, while also providing a rationale to intensify existing policies.

Note

1. It should be noted that heat pumps are not the only option available for reducing the CO_2 emissions of supplying heat. Expanding the use of district heating may also reduce CO_2 emissions as fuels can be used more efficiently (for example, by using the waste heat of industrial processes or power plants).

References

Adamantiades, A. & Kessides, I. (2009) 'Nuclear Power for Sustainable Development: Current Status and Future Prospects', *Energy Policy, 37*, 5149–5166.

Bundesregierung (2010) 'Energy Concept for an Environmentally Sound, Reliable and Affordable Energy Supply', http://www.germany.info/contentblob/3043402/Daten/3903429/BMUBMWi_Energy_Concept_DD.pdf, date accessed 5 December 2014.

Cardwell, D. (2013) 'On Rooftops, a Rival for Utilities', *The New York Times,* 26 July 2013, http://www.nytimes.com/2013/07/27/business/energy-environment/utilities-confront-fresh-threat-do-it-yourself-power.html?pagewanted=all&_r=0, date accessed 3 July 2014.

Cramton, P. & Ockenfels, A. (2011) 'Economics and Design of Capacity Markets for the Power Sector', *Zeitschrift für Energiewirtschaft, 36*(2), 113–134.

Denholm, P. & Hand, M. (2011) 'Grid Flexibility and Storage Required to Achieve Very High Penetration of Variable Renewable Electricity', *Energy Policy, 39*, 1817–1830.

Department of Energy & Climate Change (2012) *Demand Side Response in the Domestic Sector – A Literature Review of Major Trials*, Final Report Undertaken by Frontier Economics and Sustainability First (London) https://www.gov.uk/government/uploads/system/uploads/attachment_data/file/48552/

5756-demand-side-response-in-the-domestic-sector-a-lit.pdf, date accessed 3 July 2014.

Department of Energy & Climate Change (2013) 'Initial Agreement Reached on New Nuclear Power Station at Hinkley', https://www.gov.uk/government/news/initial-agreement-reached-on-new-nuclear-power-station-at-hinkley, date accessed 3 July 2014.

ECF (2010) 'Roadmap 2050 – A Practical Guide to a Prosperous, Low-Carbon Europe, Volume 1' (European Climate Foundation) http://www.roadmap2050.eu/attachments/files/Volume1_fullreport_PressPack.pdf, date accessed 3 July 2014.

EEA (2014) 'Annual European Union Greenhouse Gas Inventory 1990–2012 and Inventory Report 2014. Submission to the UNFCCC Secretariat', *EEA Technical Report*, No 9/2014 (Copenhagen: European Environment Agency).

Eurelectric (2009) 'A Declaration by European Electricity Sector Chief Executives', http://www.eurelectric.org/CEO/CEODeclaration.asp, date accessed 3 July 2014.

Eurelectric (2011) *Flexible Generation: Backing Up Renewables, Renewables Action Plan (RESAP)* (Brussels: Eurelectric), http://www.eurelectric.org/media/61388/flexibility_report_final-2011-102-0003-01-e.pdf, date accessed 3 July 2014.

Eurelectric (2013) *Power Choices Reloaded: Europe's Lost Decade? Key Messages* (Brussels: Eurelectric), http://www.eurelectric.org/media/79057/power_choices_2013_final-2013-030-0353-01-e.pdf, date accessed 3 July 2014.

Eurobarometer (2010) *Europeans and Nuclear Safety. Special Eurobarometer 324* (Brussels).

Eurobarometer (2014) *Climate Change. Special Eurobarometer 409* (Brussels).

European Commission (2011a) 'A Roadmap for Moving to a Competitive Low Carbon Economy in 2050', COM(2011) 112.

European Commission (2011b) 'Energy Roadmap 2050', COM(2011) 885 final.

European Commission (2011c) 'Energy Roadmap 2050 – Impact Assessment, Part 1/2', SEC(2011) 1565/2.

European Commission (2011d) 'Energy Roadmap 2050 – Impact Assessment, Part 2/2', SEC(2011) 1565.

European Commission (2014a) 'EU Energy, Transport and GHG Emissions. Trends to 2050. Reference Scenario 2013', http://ec.europa.eu/energy/observatory/trends_2030/doc/trends_to_2050_update_2013.pdf, date accessed 3 July 2014.

European Commission (2014b) 'Energy Infrastructure: What Do We Want to Achieve?', http://ec.europa.eu/energy/infrastructure/index_en.htm, date accessed 3 July 2014.

European Commission (2014d) 'Impact Assessment Accompanying the Communication: A Policy Framework for Climate and Energy in the Period from 2020 up to 2030', SWD(2014) 15.

Eurostat (2014) 'Energy Statistics Online Database', http://epp.eurostat.ec.europa.eu/portal/page/portal/energy/data/database, date accessed 3 July 2014.

Fraunhofer ISI (2011) 'Tangible Ways towards Climate Protection in the European Union (EU Long-term Scenarios 2050)', http://www.isi.fraunhofer.de/isi-wAssets/docs/x/de/publikationen/Final_Report_EU-Long-term-scenarios-2050_FINAL.pdf, date accessed 3 July 2014.

Global CCS Institute (2014) 'The Global Status of CCS, February 2014', http://cdn.globalccsinstitute.com/sites/default/files/publications/121016/ global-status-ccs-february-2014.pdf, date accessed 3 July 2014.

Greenpeace & EREC (2012) *Energy [R]evolution – A Sustainable EU 27 Energy Outlook* (Brussels: Greenpeace International & European Renewable Energy Council)

Grubler, A. (2010) 'The Costs of the French Nuclear Scale-up: A Case of Negative Learning by Doing', *Energy Policy, 38*, 5174–5188.

Haas, R., Resch, G., Panzer, C., Busch, S., Ragwitz, M. & Held, A. (2011) 'Efficiency and Effectiveness of Promotion Systems for Electricity Generation from Renewable Energy Sources – Lessons from EU Countries', *Energy, 36*, 2186–2193.

IEA (2012) *Energy Technology Perspectives 2012 – Pathways to a Clean Energy System* (Paris: International Energy Agency/OECD).

IEA (2014) *The Power of Transformation. Wind, Sun and the Economics of Flexible Power Systems* (Paris: International Energy Agency/OECD).

Interagency Working Group (2013) 'Technical Update of the Social Cost of Carbon for Regulatory Impact Analysis – Under Executive Order 12866', *Interagency Working Group on Social Cost of Carbon, United States Government,* http://www.whitehouse.gov/sites/default/files/omb/inforeg/social_cost_ of_carbon_for_ria_2013_update.pdf, date accessed 3 July 2014.

Joint Research Centre (2011) '2011 Technology Map of the European Strategic Energy Technology Plan (SET-Plan). Technology Descriptions', *JRC Scientific and Technical Reports,* http://setis.ec.europa.eu/system/files/Technology_Map_2011. pdf, date accessed 3 July 2014.

Kimura, O. (2010) 'Japanese Top Runner Approach for Energy Efficiency Standards', *SERC Discussion Paper,* SERC09035, http://www.climatepolicy.jp/thesis/ pdf/09035dp.pdf, date accessed 3 July 2014.

Kley, F., Wietschel, M. & Dallinger, D. (2010) 'Evaluation of European Electric Vehicle Support Schemes', *Working Paper, Sustainability and Innovation,* No. S7/2010, http://www.econstor.eu/bitstream/10419/40019/1/634898620. pdf?origin=publication_detail, date accessed 3 July 2014.

Leach, M., Rockström, J., Rasin, P., Scoones, I., Stirling A. C., Smith, A., Thompson, J., Millstone, E., Ely, A., Arond, E., Folke, C. & Olsson, P. (2012) 'Transforming Innovation for Sustainability', *Ecology and Society, 17*(2), article 11.

Lechtenböhmer, S. & Luhmann, H.-J. (2013) 'Decarbonization and Regulation of Germany's Electricity System after Fukushima', *Climate Policy, 13*(Special Issue Supplement 1: Low carbon drivers for a sustainable world), 146–154.

Mock, P. & Yang, Z. (2014) 'Driving Electrification – A Global Comparison of Fiscal Incentive Policy for Electric Vehicles', http://www.theicct.org/ sites/default/files/publications/ICCT_EV-fiscal-incentives_20140506.pdf, date accessed 3 July 2014.

Nagl, S. (2013) 'Prices vs. Quantities: Incentives for Renewable Power Generation – Numerical Analysis for the European Power Market', *EWI Working Paper,* No 13/04.

OECD (2010) *Public Attitudes to Nuclear Power* (Paris: OECD).

Owen, A. D. (2006) 'Renewable Energy: Externality Costs as Market Barriers', *Energy Policy, 34*(5), 632–642.

Prognos (2014) *Comparing the Cost of Low-Carbon Technologies: What Is the Cheapest Option? An Analysis of New Wind, Solar, Nuclear and CCS Based on Current Support Schemes in the UK and Germany* (Berlin: Agora Energiewende).

REN21 (2014) *Renewables 2014: Global Status Report* (Paris: REN21 Secretariat).

Schneider, M. & Froggat, A. (2013) *World Nuclear Industry Status Report 2013* (Paris, London: Mycle Schneider Consulting).

Seyfang G. & Haxeltine, A. (2012) 'Growing Grassroots Innovations: Exploring the Role of Community-based Initiatives in Governing Sustainable Energy Transitions', *Environment and Planning C: Government and Policy, 30*(3), 381–400.

Sweco Energuide AB (2014) 'Capacity Markets in Europe: Impacts on Trade and Investments. A Sweco Multiclient Study', *Sweco Energuide AB,* Project No. 5467470000, http://www.elforsk.se/Documents/Market%20Design/conference/2014%20Papers/3_4_report.pdf, date accessed 3 July 2014.

Vasconcelos, J., Ruester, S., He, X., Chong, E. & Glachant, J.-M. (2012) 'Electricity Storage: How to Facilitate Its Deployment and Operation in the EU', *Think Project,* Final Report June 2012, http://www.eui.eu/Projects/THINK/Documents/Thinktopic/THINKTopic8online.pdf, date accessed 3 July 2014.

Viebahn, P., Vallentin, D. & Höller, S. (2012) 'Integrated Assessment of Carbon Capture and Storage (CCS) in the German Power Sector and Comparison with the Deployment of Renewable Energies', *Applied Energy, 97,* 238–248.

von Hirschhausen, C., Herold, J. & Oei, P.-Y. (2012) 'How a "Low Carbon" Innovation Can Fail – Tales from a "Lost Decade" for Carbon Capture, Transport, and Sequestration (CCTS)', *Economics of Energy & Environmental Policy, 1*(2), 115–123.

4
Electricity Grids: No Decarbonization without Infrastructure

Thomas Sattich

Introduction

This chapter discusses the role of electricity transmission infrastructure for the integration of renewables into the European power system in the context of the EU's decarbonization goals. Conventional plants largely define today's power system, including its technical and economic environment. The integration of renewables into this system constitutes a major challenge that needs to be addressed for successful decarbonization. On the basis of current technologies, the European power system has to be significantly adjusted in order to absorb increasing amounts of renewable energy (RE). A major part of the necessary adjustments concerns the power transmission infrastructure – or grid.

The EU's decarbonization objective to reduce greenhouse gas (GHG) emissions by 80–95 per cent by 2050 implies a major transformation of Europe's energy sector towards the large-scale use of renewables. This transition from carbon-based energy towards RE has already started (see Figure 4.1). The power sector plays a crucial role in this context (see Chapter 3). It represented about 37 per cent of total CO_2 emissions in Europe in 2012 and is one of the sectors where the transformation could take place in the quickest and most economical way (Roques, 2014, p. 82). The contribution of carbon-neutral RE sources to electricity consumption grew by about 64 per cent between 2004 and 2012, so that renewables in 2012 already accounted for about 24 per cent of electricity generation (Eurostat, 2014). And with expected electrification of other sectors, including industry, transport and buildings (Sugiyama, 2012), the power sector's contribution to decarbonization is likely to grow further.

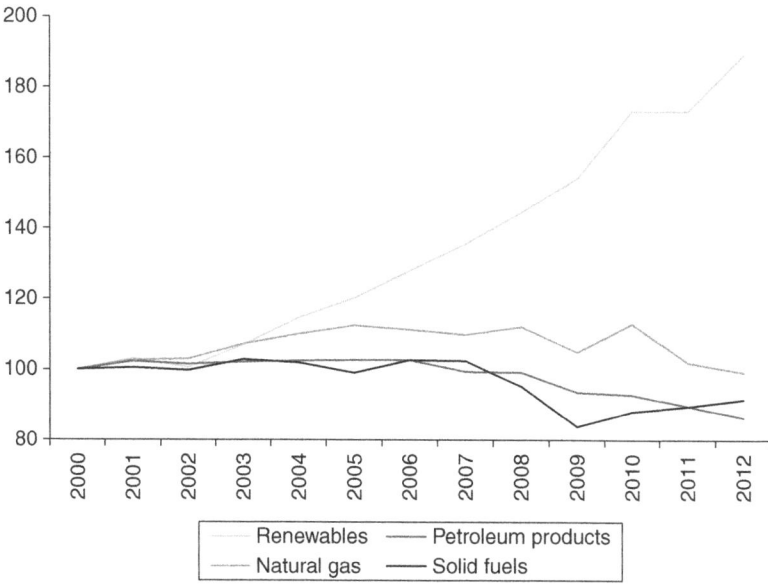

Figure 4.1 Development of different energy carriers in the EU-28 from 2000 to 2012

Note: EU gross inland energy consumption by fuel type in 1,000 tonnes of oil equivalent; 2000 = 100 per cent.

Source: Eurostat (2014).

Renewables will deliver a main part of the energy supply required to 2050 (see Chapters 1 and 3). While decarbonization is ongoing, the rapid growth of renewables and the relative decline of conventional energy sources will face a number of serious constraints (European Commission, 2013a, p. 2). Power grids play a crucial role in this context, as they are a prerequisite for flexible, interactive operation of power plants, for the efficient allocation of generation units over a given territory, and for the interconnection between generation and consumption centres. A fundamental change in power generation affects this system.

This chapter explores the challenges to Europe's power grids in the transition to a decarbonized, renewables-based power system and assesses related EU policies. The next section examines the complex relationship between renewables and the grid infrastructure. To identify (potential) gaps in the EU's approach to grid development, I then look at relevant EU programmes. The following section turns to the factors that may explain the current state and development of the

EU's power transmission infrastructure. Finally, I close the chapter by highlighting the main issues for grid development to 2050 and by providing a number of recommendations to overcome the policy challenges.

1. Power grids and decarbonization

Because storage of electricity is difficult and costly, power generation has to adapt to changing demand ('load') in real time in order to keep the grid stable. Power plants therefore operate as interacting components in integrated power pools (power generation, grids and consumption/markets) where the different generation units adjust their power output to the momentary load. Electricity grids must optimize this system, with interregional power lines providing system operators with the flexibility needed to keep the network stable despite local load changes (ECF, 2010, p. 70). The larger, more flexible and diverse a power pool is, the better a network can be stabilized.

To achieve decarbonization, European power grids need to be adapted to increasing shares of renewables. Conventional plants still play a dominant role in the structure of the power grids. The technical features of renewables, however, are different from those of conventional plants, and thus disturb the interactions in the established system (Schaber et al., 2012a, p. 123). Decarbonization of the power sector is thus more than replacing old conventional power plants with new carbon-neutral power generation units, but requires the reorganization of the environment in which renewables operate (Schaber et al., 2012b). Several studies highlight the importance of power transmission infrastructure in transitioning to renewables (see, for example, Tröster et al., 2011). Without suitable adaptations, decarbonization will at best be much more difficult to achieve and/or delayed while waiting for other technological options such as storage. Intelligent systems to predict loads, smart grids and a densely intermeshed electricity transmission grid are widely believed to be necessary elements of the adaptation (Capros et al., 2012, p. 96; Fürsch et al., 2013, p. 642).

This poses several technical, economic and political questions. What infrastructure upgrades are needed? Can infrastructure development keep pace with decarbonization requirements? What sort of drivers and barriers exist? Are relevant EU policies effective and sufficient? To answer these questions, I discuss the complex relationship between the integration of renewables and (cross-border) power transmission infrastructure, and the corresponding necessary reorganization of the European grid.

1.1 Electricity transmission networks and decarbonization

An estimated 80 per cent of bottlenecks in the European power system are related to the integration of renewables (ENTSO-E, 2012, p. 56). The pressure that RE power generation puts on the established system varies, however, with the specific type and the location of plants. Measures to upgrade the grid will therefore have to differ from region to region. The question is where exactly the increase of renewables will require adaptations and what kind of adaptations will be necessary.

Decarbonization thus poses a double challenge for the European power transmission infrastructure. First, the switch to renewables will change the distribution of power plants and hence the topography of the grid. The relatively low energy density of many RE sources will require installations such as wind parks to be dispersed over large territories, which requires interconnections. Furthermore, RE production (for example, offshore wind power installations) is likely to be far from the centres of consumption. RE sources, such as wind and sunshine, are distributed unequally, which will influence the distribution of RE production. For example, load hours of solar and wind power vary by up to 100 per cent between the most and least favourable sites in Europe (Fürsch et al., 2013, p. 650). Technological advances in RE may mean that these differences become less significant over time. Nevertheless, power transmission infrastructure will have to cover growing distances in order to bring the electricity to the consumer.

Moreover, there are differences between the various forms of renewables. While wind, solar and tidal power will be highly concentrated in areas far away from consumption centres, biomass can (partly) be transported to and used by the power generation infrastructure in place. The impact on the grid will thus vary greatly between regions and depend on the regional energy mix.

Second, the intermittency of RE production will have strong implications for the future shape of the European power grid. In contrast to traditional fossil fuel-based systems, many RE plants depend on changing elemental forces such as wind and sun. Grid modernization will thus be required to stabilize power supply.

According to the European Commission, existing power pools are flexible enough to counterbalance about 5 per cent of intermittent renewables such as wind and solar power (European Commission, 2012, p. 8). Where, however, intermittent RE greatly exceeds this margin, additional measures are required to provide the system with enough flexibility to absorb network fluctuations.

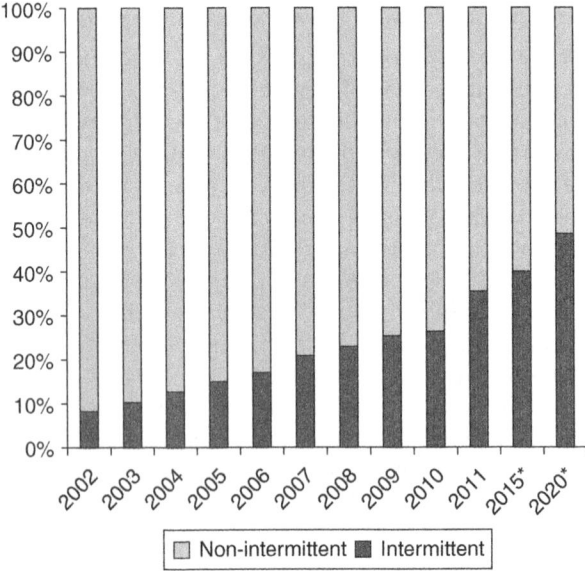

Figure 4.2   Share of intermittent and non-intermittent renewables in the EU-27 (per cent of all renewable-generated electricity)
Note: Data for 2015 and 2020 are estimates.
Source: Eurostat (2014); ECN (2011b, p. 14).

In this regard, the European power system is close to a paradigm shift. In 2011, intermittent renewables already amounted to 35 per cent of RE-generated electricity (9 per cent in 2002) and 7 per cent of all electricity generated (Eurostat, 2014). Intermittent RE is expected to climb to 49.7 per cent of RE capacity by 2020 (ECN, 2011b, p. 14; see Figure 4.2)[1] accounting for 17–20 per cent of European electricity consumption by 2020 (ECN, 2011a). The EU's power sector is thus headed towards an era where the characteristics of intermittent renewables will increasingly determine the logic of power generation, transmission and consumption.

Increasing the flexibility of existing power pools and their transformation to a smart and flexible system is thus the key to the sector's further decarbonization, and – in the long run – to a complete switch to renewables. Since the capacity to keep growing network fluctuations under control can be increased by integrating different regions, the power grid has a vital role to play in this adaptation. The need for more flexibility in particular areas depends, however, on the specific regional energy mix.

1.2 Reorganizing the European power grid

For decarbonization, the reorganization of power grids needs to focus on two aspects related to the inevitable growth of RE:

1. New infrastructure to adapt to the power sector's changing topography;
2. Adaptations to compensate for growing network fluctuations caused by intermittent renewables.

This need for grid modernization is universal, as it concerns all parts (uptake, transmission, distribution, off-take, metering and so on), levels (regional, national, international) and most geographical sections of the grid. Ideas and plans for the reorganization of the European grid are hence far-reaching, and include the construction of regional (for example, the North Seas Countries' Offshore Grid Initiative, see NSCOGI, 2010), continental (the super-grid, see FOSG, 2013) and even transcontinental networks (between North Africa and Europe, see, for example, DESERTEC, 2009). The characteristics of different RE technologies do pose specific challenges for particular sections of the power grid. The reorganization should aim to create a flexible and smart power pool of European size, where all power plants jointly balance fluctuations in power production (Battaglini et al., 2009). This constitutes a convincing and economically viable possibility for better network stabilization (Newbery et al., 2013, pp. 86–87).

Cross-border transmission infrastructure is in particular need of further development. Exchange between the different national systems stagnates at 7–10 per cent (ENTSO-E, 2014), with most member states being largely self-sufficient in electricity supply (Eikeland, 2011, p. 14). Today's power transmission infrastructure rather prevents free flow of electricity in Europe, thus preventing the rational distribution of RE generation units and the ability to balance network fluctuations (ENTSO-E, 2012, p. 56). Filling the gaps in the European cross-border grid infrastructure represents a largely untapped potential (ENTSO-E, 2012) for near complete decarbonization at moderate costs (Haller et al., 2012, p. 288).

Overall, the power system in Europe can be described as a heterogenic patchwork of more or less integrated components of still largely national power systems. In some regions, the power transmission infrastructure has significant exchange capacity, which in turn allows high levels of renewables in the system (Sattich, 2014b). Examples include the Scandinavian countries (particularly Denmark) and the German-Dutch couple. Other regions show, however, much lower cross-border power exchange capacities. The most prominent example is the French-Spanish border, where a possible significant increase of capacity for electricity exchange has not yet been achieved as of 2014.[2] In between these extremes, levels of interconnection vary but are relatively low. For example, large cross-border flows of electricity from wind power generated in Northern Germany threaten to overcharge Polish and Czech grids. In this particular case, power flows are not only erratic and unplanned, but also unwelcome by the receiving side, which is planning to block them (PSE, 2011).

Where are grid developments and the necessary adaptations most urgently required? Considering the per-area generation of intermittent electricity is helpful in this regard. Figure 4.3 shows sharp differences among member states in their approach to renewables. While some countries have only a moderate density of intermittent renewables and intend to keep their numbers limited, others are integrating large and increasing amounts into their national systems (ECN, 2011a). As a result, the need for interconnectors with neighbouring countries varies widely.

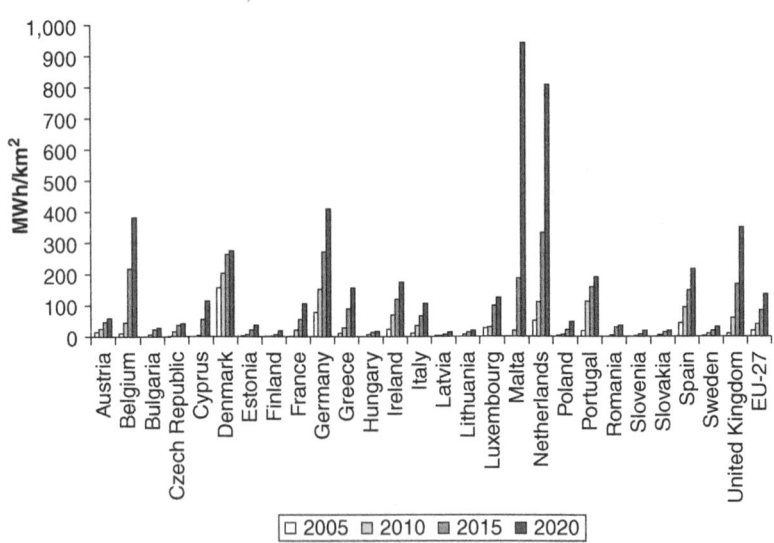

Figure 4.3 Calculated per area generation of intermittent electricity (solar, tidal, wind) (MWh/km^2)

Note: Data for 2015 and 2020 are estimates.

Source: ECN (2011a, pp. 124, 132, 142).

2. EU policy to develop the European electricity grid

The development of energy systems technology in the EU has not kept up with the growing ambitions for sustainability and decarbonization, and the deployment of new electricity transmission infrastructure has moved forward at an insufficient pace. In response, the EU has implemented a number of programmes to accelerate grid development. To determine whether these measures are sufficient to achieve decarbonization, this section first analyses past and present EU policies. I then address programmes for universal grid development and instruments for the development of specific parts of the European network. The section concludes with a discussion of the overall approach of the EU towards universal and specific grid development.

2.1   From market integration to decarbonization

The EU (then, the European Community) began pursuing deeper integration of power markets and the development of a 'common carrier system' for electricity with the Single Market Programme in the 1980s (European Commission, 1988, p. 72). The need arose from the historical, gradual, bottom-up approach to the development of cross-border power transmission infrastructure in Europe that had emerged, despite a discussion as early as the 1920s about the possibility of a top-down, supranational approach (Lagendijk, 2008, pp. 80ff; van der Vleuten & Lagendijk, 2010, p. 2045). In 1988, the European Commission proposed declaring certain large-scale energy infrastructures as being of 'Community interest' and hence entitled to special treatment (European Commission, 1988). Since then, a number of policy initiatives aimed to develop a more integrated power transmission infrastructure.

In the 1990s, power transmission capacity was found to be insufficient, while mechanisms for the support of renewables focused on the increase of RE generation capacity. As ambitions in climate policy and the deployment of renewables grew, the relationship between these trends and the goal of creating the internal electricity market was increasingly discussed (Glachant et al., 2010; see Chapter 2). The Commission concluded that power transmission capacity was indeed insufficient (Eikeland, 2011, p. 20), but existing support mechanisms largely neglected the environment in which renewables operated (see Fouquet & Johansson, 2008). Only with regard to the most obvious cases, such as distant wind parks, was the need for the adaptation of power networks and the additional costs for the installation and/or operation of renewables considered. Given the low penetration of intermittent renewables in the power system at that time, the main concern was the free mandatory access of renewables to the grid at fair prices (European Commission, 1997, pp. 14, 29; 1998, p. 8).

In 1997, the EU agreed to obtain 12 per cent of energy from renewable sources by 2010 (EWIS, 2010, p. 146), and since then, the promotion of RE has moved gradually up the European agenda (Nilsson et al., 2009, p. 4454). The importance of the power grid for the operation of renewables became more widely recognized (European Commission, 2000, p. 48). Questions concerning the electricity transmission infrastructure, conditions for grid access, grid reinforcement and charges to RE generators for the use of networks, received more attention (Jansen & Uyterlinde, 2004, p. 93). Provisions on grid access and operation nearly doubled in length from Directive 2001/77/EC on the promotion of renewable electricity (Article 7) to Directive 2009/28/EC on the promotion of renewable energy (Article 16).

2.2   Instruments for universal grid development

Relevant EU policies and programmes can roughly be categorized into two groups:

(1) top-down instruments that pull (demand-pull) and
(2) bottom-up instruments that push industry towards investments and technology development (technology push) (Sattich, 2014a).

With regard to demand-pull, the EU determined full unbundling of network and generation/supply interests in its Electricity Directive in 2009 (see Chapter 2; Eikeland, 2011, p. 32). This provides stimuli for grid development in a European rather than national or regional context. Unbundling is the most noted, but not the only European policy to support the development of the power grid.

The two Directives on the promotion of renewables (Directives 2001/77/EC and 2009/28/EC) have also been important for stimulating demand for new infrastructure. Directive 2001/77/EC determined that EU members *may* provide priority access to the grid for renewables and requested member states to implement cost sharing mechanisms for the necessary grid reinforcements (Article 7; see also EWIS, 2010, p. 146; Jansen & Uyterlinde, 2004, p. 99). Directive 2009/28/EC *requires* member states to give priority to renewable electricity and related installations. Transmission system operators should ensure that appropriate grid measures are taken to minimize the curtailment of electricity produced from

RE sources (Article 16(2); EWIS, 2010, p. 154). The 2009 Directive hence marks a milestone for the commitment to grid development.

Regarding technology push, the EU has intensified its attempts to foster the development of new energy technologies through R&D programmes. These have focused on (market) barriers that hamper the development of new and RE technology, and the transition from demonstration projects to marketing. With a multiannual budget of 200 million EUR, the Intelligent Energy Programme (Decision No 1230/2003/EC) was rather small and supported only a limited number of grid projects.[3] The EU provided more financial support for selected energy technologies under its Seventh Framework Programme for Research and Technological Development (Decision No 1982/2006/EC), cohesion funds, and other funds. The research programme Horizon 2020 (2014–2020) has an increased budget for energy-related projects and the EU Institute for Energy and Transport provides its own R&D infrastructure.

In particular, the Strategic Energy Technology Plan (SET-Plan, see European Commission, 2006b) brings market uptake measures and support for basic research together in one encompassing programme (European Commission, 2006a, p. 5). This SET-Plan aims to reinforce international cooperation and coherence among national, European and international energy research programmes (European Commission, 2007b, pp. 9–11). One of SET's subcomponents – the European Electricity Grid Initiative – aims to create an integrated R&D and demonstration network to develop, demonstrate and validate technologies for the transmission and distribution of up to 35 per cent of electricity from renewable sources by 2020, and to completely decarbonize electricity production by 2050 (European Union, 2010, p. 4). With an estimated multiannual budget of 2 billion EUR, the Grid Initiative is considerably bigger than other relevant European instruments. The implementation of the SET-Plan has, however, fallen below expectations (due to poor coordination, financing and commitment from member states) (European Commission, 2013b, pp. 8–9; SETIS, 2014).

2.3 Instruments for specific grid-development projects

The trans-European energy network programme (TEN-E) constitutes the main EU-level instrument for grid development. It emerged from debates about the need for new cross-border electricity exchange capacity in specific bottlenecks of the European grid (European Commission, 1988, p. 28), and the endorsement of a limited number of high priority trans-European projects (European Council, 1994). TEN-E aims to create favourable framework conditions for grid development in selected parts of the European network, with a strong focus on market integration. Cross-border power interconnection projects that contribute to the integration of renewables are also included (Decision No 1254/96/EC; Regulation No 347/2013, Article 4 and Annex IV).

Under the TEN-E programme, four priority electricity corridors are declared to be of 'European interest' (Regulation No 347/2013, Annex I), including the North Sea offshore grid and the North-South electricity interconnections in Central, Eastern and Southern Europe. Particular projects within these corridors such as high-voltage overhead transmission lines are labelled as being of common interest and provided with priority status that entitles them to administrative and financial support. In total, 66 power infrastructure projects fall into this category, most of which are transmission infrastructure projects (European Commission, 2013c).

Overall, TEN-E addresses the main obstacles for private sector investments in energy networks, namely administrative constraints and market conditions. One of the starting points to reduce administrative burdens is permit granting. In order to limit the process to 42 months (Regulation No 347/2013, Article 10(2)), member states are obliged to establish a competent authority responsible for all permit granting processes in a one-stop shop (Article 8(1)). Projects of common interest shall be provided with the highest national significance (Article 7). In order to facilitate coordination in the implementation of particular projects, the programme guidelines provide for close cooperation among all relevant parties (such as member states, national regulatory authorities, transmission systems operators) in regional groups (Justice and Environment, 2013, p. 3). Where a project of European interest encounters significant delays or implementation difficulties, European coordinators can initiate cross-border dialogue among the parties concerned (Regulation No 347/2013, Article 6). In addition, the TEN-E programme provides a yearly budget of 25 million EUR supporting projects that are not viable under the existing regulatory framework and the given market conditions (Regulation No 347/2013, Article 14(1)) – 20 million EUR of which is intended to co-finance feasibility studies (see Meeus et al., 2005, p. 31).

TEN-E focuses on addressing regulatory disincentives to invest (THINK, 2011, pp. 32–33) and on organizing the market for grid investments, so that investors can have confidence in the recovery of their costs (Helm, 2014, p. 31). Financial assistance must be exceptional and distortions to competition minimal, since market principles have priority. TEN-E thus provides for joint system planning and the allocation of costs and risks of new cross-border interconnections (European Commission, 2007a, p. 17). To this end, the programme guidelines foresee the development of a harmonized system wide cost-benefit analysis as the basis for the allocation of investment costs of particular projects (Regulation No 347/2013, Article 12(1)). Moreover, where the investment in new cross-border interconnectors is prevented by high risks (for example, future non-use), a number of (time-limited) exemptions provide project owners with a certain independence from the regulatory framework (Regulation No 714/2009, Article 17).

2.4   Better financing towards decarbonization?

Despite these efforts, much remains to be done to develop a power transmission infrastructure that is capable of integrating renewables (Monti, 2010, p. 48). In this respect, the aforementioned EU programmes remain insufficient (van Agt, 2011, p. 30). This may be due to two interacting factors. First, there may have been too much focus on a bottom-up approach with emphasis on specific projects (Helm, 2014). Given the largely varying need for interconnection, however, specific bottlenecks are the obvious starting point for grid adaptations. With the TEN-E programme, EU policy has moved towards a more comprehensive top-down approach (van Agt, 2011, pp. 28–29). Second, grid development also seems to have suffered from a lack of investor confidence that costs can be recovered (Helm, 2014, pp. 30–31). In response, the European Commission called for a new approach to the planning, construction and operation of electricity transmission infrastructure and set out to develop innovative funding mechanisms for the coverage of the main risks to improve the investment climate (European Commission, 2010).

The EU has thus become increasingly inclined to (co-)finance or guarantee the finance of infrastructure projects and to focus public support on projects of European interest (van Agt, 2011, p. 29). The European Energy Programme for Recovery established by Regulation No 663/2009 – in response to the economic crises of 2008 – provided financial support to selected energy infrastructure projects (European Commission, 2013b). The connection and integration of RE sources is among the five objectives of this programme. It made 904 million EUR available for grid interconnection measures and 565 million EUR for offshore wind energy – considerable sums compared to the financial resources of the TEN-E programme.[4]

The EU established the Connecting Europe Facility in 2013 (Regulation No 1316/2013) with a multiannual budget of 5.85 billion EUR to support energy infrastructure projects of common interest that contribute to smart, sustainable and inclusive growth or enable the EU to achieve its GHG emission reduction targets (Regulation 1316/2013, Articles 3 and 5). This support will mostly be implemented in the form of grants, procurement and financial instruments as risk sharing instruments for project companies (Regulation 1316/2013, Article 6).

Compared to the overall investment needs of about 600 billion EUR for (smart) grids and storage (Tagliapietra, 2013), the financial support provided by the EU can be seen as limited. Support for energy infrastructure projects and the upgrade of Europe's electricity networks for the integration of renewables are, however, prioritized in the Energy 2020 strategy. Increasing certainty for investment and innovation in grids is one of the key elements of this strategy. Energy 2020 could be expected to include further financial resources for the necessary grid integration of renewables (European Commission, 2010, p. 16). The financial scope under Energy 2020 is, however, unclear.

3. Explaining EU policy on the power grid

This section accounts for a number of drivers and barriers that may help explain the insufficient progress in the development of Europe's power transmission infrastructure, following the main factors introduced in Chapter 1.

3.1 Functional overlap

Policy to support the development of the European grid is crucial for integrating higher levels of renewably generated electricity, and for the move away from fossil fuels and towards decarbonization. Policies to support grid development for the integration of renewables are thus harmonious or in synergy with the goal of decarbonization. However, when policies to upgrade the grid enhance or perpetuate the use of fossil fuels, these policies are not necessarily in synergy. Therefore, the nature of the objective of grid upgrade is crucial for a synergetic overlap between grid policy and decarbonization goals. As outlined above, the integration of renewables has become one of the defining features of grid-development policy, albeit with insufficient results. Policy challenges persist, even if the functional overlap between grid development for the integration of renewables and decarbonization objectives is synergetic. This may be related to the interactions with multiple policy goals, as grid-development policy overlaps with other objectives as well as decarbonization objectives.

First, in the early days of grid policy development, the dominant motivation was to achieve further market integration. European grid-development policy started with the ambition to build a common electricity market, and the integration of renewables was later added as a motivation. As a result, EU programmes to integrate the electricity market saw a remarkable evolution towards a twofold end, which entails the integration of renewables in addition to market integration. Most instruments to foster grid development hence have their origins in the internal market programme, but incorporated the integration of renewables. As an add-on policy element, it is possible that the overlap between grid-development policy and the integration of renewables is overshadowed by the initial goal of market integration – perhaps rendering the synergetic functional overlap with decarbonization goals insufficient to push policy forward.

Second, considerations of the financial and competition implications for member states to support policies to develop a more integrated grid system may trump the clear co-benefits for decarbonization objectives to strengthen the grid. Grid-development policy instruments at the EU level focus on research and development and financial support mechanisms. In times of economic crisis, providing more finance to EU projects and programmes that may not immediately benefit all member states may prove difficult. This has led to a rather uncoordinated set of policy instruments at the EU-level, with varying budgets (the SET-Plan, Horizon 2020, TEN-E and the Connecting Europe Facility, for example). In contrast to their relatively strong relationship, there is little coordination among the policy instruments, and there hardly is any common structure to coordinate all elements of grid development. Centralized funding would be a prime candidate to reach better coordination than is the case in 2014 among the different instruments.

In sum, while the functional overlap between decarbonization objectives and policy measures to ensure the integration of high levels of renewables into the European grid is synergetic, competing policy overlaps also exist. Objectives to improve market integration do not necessarily focus on the better integration of renewables. Financial concerns over investments required may dampen enthusiasm for grid-development policy. Therefore, the market integration motivation and financial concerns may balance or even trump consideration of decarbonization goals.

3.2 Political commitment

Some level of commitment to upgrading the European grid in line with objectives to increase the share of renewables is evident in decisions to assign priority labels to a number of relevant projects (such as projects of 'European' or of 'common' interest). This shows recognition that grid upgrading is required, and that some level of political support for measures to enable the upgrade exists. Additionally, political commitment to combatting climate change in a general sense in the EU is evident through a stream of policy measures to reduce GHG emissions (see Chapter 1). However, the political commitment to climate policy may not translate so smoothly into political commitment for appropriate grid infrastructure development towards decarbonization, as it is tempered in the field of grid development by other political concerns.

First, even within the EU, the idea of increased energy dependence on other states may prove politically unpopular, thus reducing commitment to increased cross-border grid integration. Increasing cross-border power exchange capacity will increase the likelihood of generation capacity migrating beyond national borders. If fully implemented and effective, EU grid-development policies will thus lead to a reshuffling of market areas, and (in particular cases) increased dependency on neighbouring states to uphold electricity supply in another. Where high levels of power system integration already exist, one can infer not only high levels of mutual trust between the involved nations (as in Scandinavia), but also a distinctive political will to implement a policy that will ultimately increase dependency on others (as in the case of Germany and the Netherlands). Political will to upgrade the European grid may thus be reduced by a reluctance to implement policies that will potentially increase dependency.

Second, political concerns to protect national power generation capacity may also affect the level of political commitment to any EU grid-development plan. As interconnection opens the door to more competition between power companies, some member states and power companies may focus on the national economic implications of increased competition as a reason for slowing down grid development. On the flipside, strong integration also indicates an orientation towards the beneficial role of market forces, which may appeal to member states and lobbyists that favour higher competition, open markets and consumer rights (see also Chapter 2). The success of EU grid-development policy may therefore depend on EU policymakers' ability to persuade the parties concerned that such a policy is beneficial.

In sum, political commitment to grid development for decarbonization is rather tempered by national concerns related to energy dependency and competition. These political concerns seem to remove a sense of urgency from policy development on the European grid. Appeals to political actors to increase their commitment to grid development for the sake of decarbonization will need to take account of, and show great sensitivity towards, the interests of parties involved.

3.3 Stakeholder involvement and support

Historically the main stakeholders addressed by European grid-development policy included member states, power companies, and industries, with little room for or interest from environmental or climate advocates. The stakeholders to be addressed are much more diverse than this historical set-up, and may also be operating and located at different levels of governance (local, regional, national or EU-level). The inclusion of a wide range of stakeholders at different levels in grid-development policy is a great challenge. Yet from a decarbonization perspective, the involvement of more stakeholders, especially those that would high-light the decarbonization co-benefits of grid-development policy, would be productive. But as the case of TEN-E policy development shows, environmental advocates do not always play a role in the consultation process, both because they are not seen as important stakeholders and because these organizations themselves have not always access to policy processes on grid development (Hauser, 2011; Vasileiadou & Tuinstra, 2013).

The Ten-Year Network Development Plan is an instrument that achieves a degree of coordination among stakeholders for collective grid planning, although without highlighting the need for climate stake-holders (Regulation 714/2009, Article 8). This model could be extended and serve as an example for the integration of multiple stakeholders in policy and planning processes. The Californian Renewable Energy Transmission Initiative (RETI) is another interesting example of an open and transparent process that ensured the contribution of a diverse range of stakeholders, not only in policy development, but also in decision-making (see Olsen et al., 2012). It could prove an interesting example for the EU to ensure an open policymaking process in its future grid policies.

Besides the external stakeholders, the environmental actors within the EU institutions (DG Environment, DG Climate Action, the environment committee in the Parliament and the Council of Environment Ministers) were not necessarily much involved in grid-development policies, as energy policymakers were in the lead. Therefore, as environmental and climate EU institutional and civil society stakeholders were not promi-nently involved, making the connections to decarbonization may have been more challenging. Furthermore, stakeholders are of a very hetero-geneous nature and operate at multiple levels. From a decarbonization perspective, this may result in some of the potential co-benefits between grid development and decarbonization objectives being lost during the policy negotiations.

3.4   Institutional set-up

EU policy measures to develop a European grid capable of handling RE growth are rather a sum of disparate policy instruments than a coordinated policy field. These instruments include the internal market legislation, RE legislation, instruments to finance research, and financial mechanisms to support specific infrastructure projects. Each of these policy instruments has a specific objective, but together they do not arrive at a single, coordinated vision for EU grid development. This disparate picture may partly be explained by the institutional context within which such policies are made.

First, energy policy generally is a mixed competence of the EU. Member states retain decisions on their national energy mix at national level and grid development and operation depends on the type of energy source (intermittent or stable). Furthermore, grid development and operation may even be a regional or local competence within a member state. Grid operators consider the demand and supply with their own region and do not necessarily view the EU-level as relevant. Such competence issues mean that states may be less than eager to agree to further EU-level policy development in such a core area of state sovereignty.

Second, the policy patchwork at the EU-level leads to the further institutional challenge of ensuring that policy measures are coherent. This implies coordination horizontally among various policymakers and stakeholders at the EU-level, and coordination vertically across multiple levels of governance and implementation. As seen already in the discussion on stakeholder involvement, ensuring adequate coordination among all relevant actors at all levels is problematic even at the policy development stage. To reach better coordination among the different elements of the individual policies, more centralization, coordination and co-management may reduce organizational redundancies and result in greater synergies between different programmes. Such coordination must be encouraged, even if problematic under conditions of mixed competences. Unilateral steps in energy policy and grid development undermine efforts for a coordinated grid policy needed for decarbonization, and should hence be avoided if the EU is to achieve its decarbonization targets (Sattich, 2014b). The institutional context for grid-development policy in the EU therefore presents a challenge for a coordinated development of an integrated grid that is capable of integrating high levels of renewable energy.

Conclusions

As is clear from the discussion in this chapter, EU policies to upgrade the power grid to cater for high levels of RE generation have evolved over time, but remain insufficient for the achievement of decarbonization goals to 2050. Grid development is a prerequisite for the increased shares of renewables that are required to move to decarbonization (see also Chapter 3). This will result in a shifting topography of the power system, or in other words, a relocation of power generation capacity to new places, potentially far from centres of consumption. European decarbonization policy will thus produce winners and losers; that is regions that will lose, and others that will win new power-generation sites. Policies that sufficiently support such changes are not yet in place, as of 2014.

There are several reasons for this insufficient policy development. Historical functional links with energy market integration have promoted grid development, but not necessarily for the purpose of integrating more renewable energy. Integrating renewables is a late addition to the objectives of grid-development policy, but despite the apparent synergies there is a lack of a common vision to coordinate all elements of grid development towards decarbonization. The institutional set-up of the EU has not yet responded to the greater needs for both horizontal and vertical coordination of policy, planned projects and stakeholders. Moreover, stakeholders in this sector are not all sufficiently integrated in EU policymaking, and political commitment to grid development to achieve decarbonization is rather tempered by concerns over energy dependence, financial concerns and national economic interests.

Several strategies may alleviate some of these challenges to promoting sufficient grid-development policy at the EU-level. First, reframing the importance of decarbonization in general, and grid development in particular, as urgent responses to the needs of a growing EU industry (the renewable industry) may be helpful to focus EU policies, public support and stakeholders on the decarbonization goal. This may be especially relevant for dealing with concerns over competition in times of economic crisis. Increased involvement of different actors from all levels and branches of society in the policy development and implementation may further highlight the functional overlaps between grid development and decarbonization. Such a strategy could include inviting new actors to policy negotiations in order to underline the importance of this policy area and to increase support for, and the legitimacy of, the respective measures.

Second, supporting coordination of policy development and of stakeholders may prove essential for timely development. The discussion of positions, sensitivities and interests of the stakeholders involved in grid-development projects is a vital element for the successful adaptation of Europe's power transmission infrastructure to decarbonization. Individual projects should therefore be based on a detailed stakeholder analysis that assesses the factors discussed above and ways to alter them according to the requirements of decarbonization. Thus, an organizational structure is needed that is clear and accessible for all actors that are concerned. Centralizing (or at least better coordinating) financing for different programmes would aid coordination. A more active role for European energy regulators may help achieve more coherence with regard to integrating and coordinating all relevant stakeholders (including environmental and climate stakeholders).

Finally, as mentioned above, European energy policy to integrate power systems in Europe is set to produce winners and losers. Some member states will become electricity exporters, and others will become more dependent on their neighbours for their electricity supply. Thus, in order to make it from policy papers to reality, the integration of the European power system requires high levels of mutual trust, solidarity and political commitment to overcome the countervailing economic antagonisms.

Notes

1. The potential of hydropower, the most important non-intermittent form of renewables, is already realized. The relative share of this form of RE will therefore decline with the rise of other forms of RE (see Šturc, 2012, pp. 4–5).
2. In view of existing (RE) generation overcapacities on the Iberian peninsula, more exchange capacity is one of the prerequisites for further decarbonization in this geographic area.
3. See the project database http://ec.europa.eu/energy/intelligent/in-action/ projects-stories/index_en.htm, date accessed 17 October 2014.
4. The Energy Programme for Recovery also financed fossil fuel-related infrastructure projects.

References

Battaglini, A., Lilliestam, J., Haas, A. & Patt, A. (2009) 'Development of SuperSmart Grids for a More Efficient Utilisation of Electricity from Renewable Sources', *Journal of Cleaner Production*, 17(10), 911–918.

Capros, P., Tasios, N., De Vita, A., Mantzos, L. & Paroussos, L. (2012) 'Transformations of the Energy System in the Context of the Decarbonisation of the EU Economy in the Time Horizon to 2050', *Energy Strategic Reviews*, *1*(2), 85–96.

DESERTEC (2009) *The DESERTEC Concept* (Hamburg: DESERTEC Foundation), https://dl.dropboxusercontent.com/u/2639069/DESERTEC%20Concept. pdf, date accessed 11 November 2014.

ECF (2010) *Roadmap 2050, Volume III: Graphic Narrative* (Brussels: European Climate Foundation).

ECN (2011a) *Renewable Energy Projections as Published in the National Renewable Energy Action Plans of the European Member States* (Petten: Energy Research Centre of the Netherlands & European Environment Agency).

ECN (2011b) *Renewable Energy Projections as Published in the National Renewable Energy Action Plans of the European Member States, Summary Report* (Petten: Energy Research Centre of the Netherlands & European Environment Agency).

Eikeland, P. O. (2011) 'EU Internal Energy Market Policy: Achievements and Hurdles', in V. L. Birchfeld & J. S. Duffield (eds.), *Towards a Common European Union Energy Policy: Problems, Progress, and Prospects* (New York: Palgrave Macmillan), pp. 13–40.

ENTSO-E (2012) *Ten-Year Network Development Plan 2012* (Brussels: European Network of Transmission System Operators for Electricity).

ENTSO-E (2014) 'Online Exchange Database', https://www.entsoe.eu/data/data-portal/exchange/Pages/default.aspx, date accessed 5 December 2014.

European Commission (1988) 'The Internal Energy Market', COM(88) 238.

European Commission (1997) 'Energy for the Future: Renewable Sources of Energy', COM(97) 599.

European Commission (1998) 'Report to the Council and the European Parliament on Harmonization Requirements. Directive 96/92/EC Concerning Common Rules for the Internal Market in Electricity', COM(98) 167.

European Commission (2000) 'Towards a European Strategy for the Security of Energy Supply', COM(2000) 769.

European Commission (2006a) 'Towards a European Strategic Energy Technology Plan', COM(2006) 847.

European Commission (2006b) 'A European Strategy for Sustainable, Competitive and Secure Energy', COM(2006) 105.

European Commission (2007a) 'Prospects for the Internal Gas and Electricity Market', COM(2006) 841.

European Commission (2007b) 'A European Strategic Energy Plan (SET-Plan) – Towards a Low Carbon Future', COM(2007) 723.

European Commission (2010) 'Energy 2020: A Strategy for Competitive, Sustainable and Secure Energy', COM(2010) 639.

European Commission (2012) 'Renewable Energy: A Major Player in the European Energy Market', COM(2012) 271.

European Commission (2013a) 'On the Implementation of the European Energy Programme for Recovery', COM(2013) 791.

European Commission (2013b) 'Review of the SET-Plan Implementation Mechanisms for the Period 2010–2012', http://setis.ec.europa.eu/system/files/SET-Plan_%20Review%20of%20Implementation%202010-12.pdf, date accessed 5 December 2014.

European Commission (2013c) 'Commission Delegated Regulation (EU) No .../.. of 14.10.2013 Amending Regulation (EU) No 347/2013 of the European Parliament and of the Council on Guidelines for Trans-European Energy Infrastructure as regards the Union List of Projects of Common Interest', C(2013) 6766.

European Council (1994) 'Presidency Conclusions', Bulletin of the European Communities No. 6/1994, 24 and 25 June.

European Union (2010) *The European Strategic Energy Technology Plan, SET-Plan. Towards a Low-Carbon Future* (Luxembourg: Publications Office of the European Union).

Eurostat (2014) 'Statistics Online Database', http://epp.eurostat.ec.europa.eu/portal/page/portal/eurostat/home/, date accessed 5 May 2014.

EWIS (2010) *EWIS Final Report* (Brussels: Wilhelm Winter & European Wind Integration Study).

FOSG (2013) *FOSG Statement on the EC Roadmap for Moving to Competitive Low Carbon Economy in 2050* (Brussels: Friends of the Supergrid).

Fürsch, M., Hagspiel, S., Jägemann, C., Nagl, S., Lindenberger, D. & Tröster, E. (2013) 'The Role of Grid Extensions in a Cost-efficient Transformation of the European Electricity System until 2050', *Applied Energy, 104*, 642–652.

Fouquet, D. & Johansson, T. B. (2008) 'European Renewable Energy Policy at Crossroads – Focus on Electricity Support Mechanisms,' *Energy Policy, 36*, 4079–4092.

Glachant, J.-M., Grant, R., Hafner, M. & de Jong, J. (2010) 'Toward a Smart EU Energy Policy: Rationale and 22 Recommendations', *EUI Working Paper, RSCAS, 2010*(52).

Haller, M., Ludig, S. & Bauer, N. (2012) 'Decarbonization Scenarios for the EU and MENA Power System: Considering Spatial Distribution and Short Term Dynamics of Renewable Generation', *Energy Policy, 47*, 282–290.

Hauser, H. (2011) 'European Union Lobbying Post-Lisbon: An Economic Analysis', *Berkeley Journal of International Law, 29*(2), 680–709.

Helm, D. (2014) 'The European Framework for Energy and Climate Policies,' *Energy Policy, 64*, 29–35.

Jansen, J. C. & Uyterlinde, M. A. (2004) 'A Fragmented Market on the Way to Harmonisation? EU Policy-making on Renewable Energy Promotion', *Energy for Sustainable Development, 8*(1), 93–107.

Justice and Environment (2013) *Regulation (EU) No. 347/2013 on Guidelines for Trans-European Energy Infrastructure. Legal Analysis* (Brno: Justice and Environment).

Lagendijk, V. (2008) *Electrifying Europe: The Power of Europe in the Construction of Electricity Networks* (Amsterdam: Aksant Academic Publishers).

Meeus, L., Purchala, K. & Belmans, R. (2005) 'Development of the Internal Electricity Market in Europe', *The Electricity Journal, 18*(6), 25–35.

Monti, M. (2010) 'A New Strategy for the Single Market: At the Service of Europe's Economy and Society', http://ec.europa.eu/bepa/pdf/monti_report_final_10_05_2010_en.pdf, date accessed 5 December 2014.

Newbery, D., Strbac, G., Pudjianto, D. & Noël, P. (2013) *Benefits of an Integrated European Energy Market* (Amsterdam: Booz & Company; London: LeighFischer).

Nilsson, M., Nilsson, L. J. & Ericsson, K. (2009) 'The Rise and Fall of GO Trading in European Renewable Energy Policy: The Role of Advocacy and Policy Framing', *Energy Policy, 37*, 4454–4462.

NSCOGI (2010) *The North Sea Countries' Offshore Grid Initiative: Memorandum of Understanding* (Brussels: Benelux General Secretariat).

Olsen, D., Byron, J. & DeShazo, G. (2012) 'Collaborative Transmission Planning: California's Renewable Energy Transmission Initiative', *IEEE Transactions on Sustainable Energy, 3*(4), 837–844.

PSE (2011) *Statement of PSE Operator S.A. on the Publication "Polen gefährdet deutsche Energiewende" by Spiegel Online (04.12.2011)* (Konstancin-Jeziorna: Polskie Sieci Elektroenergetyczne).

Roques, F. (2014) 'European Electricity Markets in Crisis: Diagnostic and Way Forward', in D. Auverlot, É. Beeker, G. Hossie, L. Oriol & A. Rigard-Cerison (eds.), *The Crisis of the European Electricity System: Diagnosis and Possible Ways Forward* (Paris: Commissariat général à la stratégie et à la prospective), pp. 77–117.

Sattich, T. (2014a) 'European Energy and Industrial Policy Realigned: Risk or Opportunity for EU Eco-Innovation Strategy?', *Institute for European Studies, Policy Brief*, 03/2014.

Sattich, T. (2014b) 'Germany's Energy Transition and the European Electricity Market: Mutually Beneficial?', *Journal of Energy and Power Engineering, 8*(2), 264–273.

Schaber, K., Steinke, F. & Hamacher, T. (2012a) 'Transmission Grid Extensions for the Integration of Variable Renewable Energies in Europe: Who Benefits Where?', *Energy Policy, 43*, 123–135.

Schaber, K., Steinke, F., Mühlich, P. & Hamacher, T. (2012b) 'Parametric Study of Variable Renewable Energy Integration in Europe: Advantages and Costs of Transmission Grid Extension', *Energy Policy, 42*, 498–508.

SETIS (2014) 'SET-Plan Review (2010–2012)', http://setis.ec.europa.eu/set-plan-implementation/set-plan-review-2010-2012, date accessed 11 November 2014.

Šturc, M. (2012) 'Renewable Energy: Analysis of the Latest Data on Energy from Renewable Sources', *Eurostat Statistics in Focus*, 44/2012.

Sugiyama, M. (2012) 'Climate Change and Electrification', *Energy Policy, 44*, 464–468.

Tagliapietra, S. (2013) 'Financing the EU Energy Infrastructure after the Euro Crisis', http://blogs.lse.ac.uk/eurocrisispress/2013/10/27/financing-the-eu-energy-infrastructure-after-the-euro-crisis/, date accessed 14 May 2014.

THINK (2011) 'Transition towards a Low Carbon Energy System by 2050: What Role for the EU. Final Report, June 2011', http://www.eui.eu/PROJECTS/THINK/DOCUMENTS/THINKTOPIC/THINK2050REPORT.PDF, date accessed 2 April 2014.

Tröster, E., Kuwahata, R. & Ackermann, T. (2011) *European Grid Study* (Langen: Energynautics GmbH).

van Agt, C. (2011) 'The Energy Infrastructure Challenge', in K. Barysch (ed.), *Green, Safe, Cheap: Where Next for EU Energy Policy?* (London: Centre for European Reform), pp. 27–35.

van der Vleuten, E. & Lagendijk, V. (2010) 'Transnational Infrastructure Vulnerability: The Historical Shaping of the 2006 European "Blackout"', *Energy Policy, 38*, 2042–2052.

Vasileiadou, E. & Tuinstra, W. (2013) 'Stakeholder Consultations: Mainstreaming Climate Policy in the Energy Directorate?', *Environmental Politics, 22*(3), 475–495.

5
Decarbonizing Industry in the EU: Climate, Trade and Industrial Policy Strategies

Max Åhman and Lars J. Nilsson[1]

Introduction

Decarbonizing society poses both threats and opportunities for the manufacturing industry. For the industry that manufactures end-user products, decarbonization presents a potential to innovate higher added value clean-tech products and to expand into new 'green' markets. For the energy intensive industry (EII), that produces mainly basic materials such as steel, cement, aluminium and basic plastics, the opportunities are less obvious and the challenges greater.

The EU objective to reduce GHG emissions by 80–95 per cent by 2050 relative to 1990 includes a suggested industry sector ambition of 83–87 per cent reduction (European Commission, 2011a). To reach such near-zero emissions in the EII entails major technological shifts that are still relatively unexplored. It appears much more difficult for the EII than for other sectors, not least from an economic and policy perspective (Nilsson et al., 2011; Åhman et al., 2012).

Technical options for decarbonizing energy supply, buildings and transport are better understood or developed, and these sectors and their markets are much less exposed to international competition than industry. Industry operates in markets where ownership and value chain structures are increasingly global and integrated across continents, and competition is increasing from emerging economies (European Commission, 2013a).

Parts of the manufacturing industry, for example, consumer electronics, develop products with life cycles of two to three years and sometimes even less, with corresponding retooling. In contrast, investment cycles of 20–40 years are not uncommon in the energy and capital-intensive the EU definitions used in the Energy Tax Directive (2003/96/EC), this sector accounted for about 2.1 per cent of GDP and employed directly 3.7 million people in 2004 (Bergman et al., 2007).

When analysing the prospects for decarbonizing industry it is necessary to distinguish between the EII that produces basic materials and the lighter manufacturing industry. The opportunities and challenges in a low-carbon transition differ substantially between them.

For down-stream light manufacturing that finishes or assembles intermediate and final products, the main challenge in a low-carbon transition is the ability to innovate new products and adapt to a new and future 'green' demand (for example, for energy efficiency, electric mobility and resource efficiency). Moving towards a decarbonized EU in 2050 will create market demand for new climate-friendly products associated with low-carbon energy sources, their supporting systems and infrastructure, smart grids, passive housing, electrification of transport, and energy storage technologies. A low-carbon transition will thus provide many opportunities for innovative firms to develop and expand into new markets. This sector can more easily pass through potential cost increases from carbon neutral energy supply since energy is typically a small share of production costs.

Down-stream manufacturing requires high-quality basic materials from the up-stream and energy intensive part of the value chain that produces materials such as cement, steel, aluminium, organic chemicals (such as ethylene for making polyethylene) and nitrogen fertilizer. In this sector, the challenges of decarbonization are greater since energy constitutes a considerable share of production costs and major changes in basic process technologies may be required. Petrochemicals need to be replaced with chemicals from other feedstock. When going beyond marginal emission reductions, there seems to be relatively few, if any, co-benefits from decarbonization.

Reducing emissions to near zero in this industry sector, through carbon capture and storage (CCS), or switching to non-fossil fuels and feedstocks, is likely to result in substantially higher production costs. Higher prices for metals and other basic materials are unlikely to be a problem for the economy. The EII accounts for only a small share of GDP and the cost-share of basic materials in finished products is generally small, indicating that cost increases can be absorbed. The problem is that, all else being equal, high production costs threaten international competitiveness and may lead to carbon leakage (that is, when production is relocated to countries with less costly climate policies).

industries and technical lifetimes for equipment may extend be:
that (Lempert et al., 2002). Thus, investment decisions and stra
development efforts made today partly determine the options and r
for manoeuvre for industrial emission reduction by 2050.

Industry in Europe is increasingly recognizing the need for a
carbon transition. Sector roadmaps have begun to explore the challe
therein. But there is still considerable uncertainty concerning how g
can be reached and a transition governed. The long-term risk of ca
leakage in an EU decarbonization scenario, and the capital intensity
other characteristics of the EII, force policymakers to consider two (
tested and difficult issue areas: trade barriers through import tariff
carbon border taxes and targeted industrial development policies.

In this chapter we analyse the challenges for decarbonizing indu:
with a focus on the capital and energy intensive basic materials inc
try. The next section provides a brief overview of the industrial sec
its energy use and emissions. The following section reviews the tech
ogy options and potential implications for the overall energy syster
decarbonizing industry. Next, we describe and analyse the current
icy landscape of energy, climate, trade and industrial policy. We
assess the functional overlap, political will, societal interest and cu
institutional set-up, which provides the basis for finally discussing
is needed for an integrated policy strategy.

1. Manufacturing and the energy intensive industry

Industry is an important part of the climate problem because of its
sions. It is also an important part of the solution since it will produ
low-carbon technologies that are needed to decarbonize in all se
The manufacturing industry in the EU accounts for more than
cent of GDP; consists of more than 2 million enterprises; and en
more than 30 million people (European Commission, 2013a).

Manufacturing begins with materials that are transformed in d
complex and geographically dispersed value chains until reach
end-use market and thereafter the end-of-life market (waste/rec)
The industrial sector is, to various degrees, integrated across th
chains from primary materials to end-use products. It include
industries, such as specialized manufacturers of high-tech produ(
high value added and relatively low emissions such as pharn
cals and electronics. It also includes the EII, including prodi
basic materials with high energy and carbon emissions intensi
as steel and cement. The definition of the EII varies, but acco

The EII may seem unimportant from an economic and employment point of view (Neuhoff et al., 2014), but materials are essential to the economy and for reducing emissions in other sectors. The possibilities for substitution by less energy intensive materials are often limited and substitution through imports leads to carbon leakage. An argument forwarded by industry for maintaining production capacity of basic materials in Europe, in addition to employment and carbon leakage, is that the geographical proximity between different stages of the value chain is important for innovation in materials and product development. The importance of this link between low and high added value industries and innovation has some theoretical and scientific support (Boschma, 2005; Hansen & Winther, 2011).

1.1 Energy use and emissions in the EU industrial sector

Industry used 3,370 Terawatt-hours (TWh) of energy in 2010, equivalent to about 22 per cent of total final energy use in the EU. The major share of energy used is electricity (30 per cent) and natural gas (29 per cent), with smaller shares of renewables (7 per cent), oil products (11 per cent) and coal/coke (13 per cent). The energy use is concentrated in a limited number of EIIs that account for a major share of industrial energy use (about 70 per cent) and emissions (Lapillone et al., 2012). In addition to this, industry uses another 1,190 TWh of fossil fuels for 'non-energy consumption' – mainly naphtha, natural gas and other petroleum products as feedstock for producing, for example, plastics and nitrogen fertilizer.

Industry emissions directly account for about 20 per cent of total CO_2 emissions in the EU. Emissions from industry originate from the combustion of fossil fuels, emissions from the process itself and indirect emissions from the use of electricity (usually reported as power sector emissions). Process emissions occur during the production of cement (converting calcium carbonate to calcium oxide), iron (which involves carbon monoxide to reduce iron oxide to elemental iron) and aluminium (where electrolytic reduction of aluminium oxide consumes the carbon anodes). Combustion of fossil energy is the main source of CO_2 emissions, but more than 25 per cent are process emissions (see Figure 5.1).

Total CO_2 emissions from industry have declined since 1990 (Figure 5.1). The reductions are mainly attributed to increasing energy efficiency and fuel shifts (Lapillone et al., 2012). As shown in Figure 5.1, the process emissions that are directly linked to production volumes have not decreased much. There have been some changes in activity

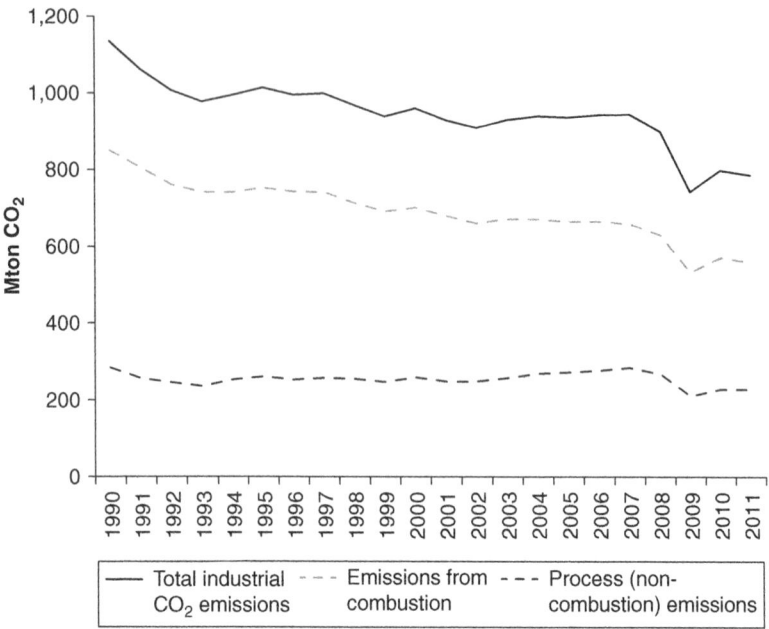

Figure 5.1   Industrial CO_2 emissions in the EU-28
Note: Not including indirect emissions from purchased electricity or other energy carriers.
Source: Adapted from European Commission (2013b).

(that is the number of tonnes produced) and structure (the relative share of products) that have reduced process emissions but these changes have only had a limited impact on the overall trend (Lapillone et al., 2012).

Long-term forecasts for basic materials show relatively stable production levels to 2040 and beyond (IEA, 2013). New production capacity will mainly be added in developing countries where demand for basic materials is growing due to industrialization. Demand for some materials, for example, cement and steel, appeared to level out in China in 2013, but production is expected to continue to increase in India (IEA, 2013).

2. Towards decarbonization of industry

In 2011, the European Commission published a 'Roadmap for moving to a competitive low-carbon economy in 2050' (European Commission, 2011a) combined with an 'Energy Roadmap 2050' with a focus on the energy sector (European Commission, 2011b; see also Chapter 1). Since mitigation options and challenges are specific for each industry sector, the European Commission expressed the need to develop roadmaps in cooperation with the industrial sectors concerned. Since then, and encouraged by DG Enterprise, industry associations in the EU have developed their own roadmaps to present their views on the technical opportunities and challenges as well as the policy implications. So far the paper industries (CEPI, 2011), chemical industry (CEFIC, 2013), steel association (Eurofer, 2013), glass association (Glass for Europe, 2013), cement association (CEMBUREAU, 2013) and the aluminium association (EAA, 2012) have published their own roadmaps for 2050.

In the short to medium term, the main mitigation options are to improve energy efficiency and to shift to less carbon intensive fuels, for example, from coal to natural gas and from natural gas to biofuels. In the IPCC Fifth Assessment Report, it is estimated that energy intensity in industry can be reduced by up to 25 per cent through wide-scale deployment of best available technologies (IPCC, 2014). We estimate that the short- to medium-term mitigation potentials range from 10 to 30 per cent for EII. This estimate is based on Åhman et al. (2012), the IPCC report (2014) and the industrial roadmaps listed above. It includes energy efficiency and fuel shifts that require adoption of best available technologies, often with co-benefits, but no fundamental changes to existing core processes.

The Commission and the industrial roadmaps are less detailed on the long-term technical options and provide only rough sketches of the development post-2030. A common argument from industry is that emission reductions beyond 50 per cent require development of 'breakthrough technologies', including solutions that are neither technically mature today nor cost-competitive without a high price on CO_2 emissions. CCS is put forward as a key back-stop[2] technology in both the European Commission's and the industry's roadmaps, despite uncertainty concerning the feasibility of CCS (see below). Compared to the Commission's roadmaps, the industrial roadmaps recognize and provide more examples of the non-CCS options (such as electrification and new materials), and most of these involve major shifts in core process technologies.

2.1 Long-term technology options

The performance, costs and other characteristics of low-carbon technologies are important for understanding the prospects for a new institutional set-up. In addition to CCS, the potential energy carriers and energy sources to replace fossil fuels in the EII are limited to bioenergy and electricity from renewables or nuclear energy. For decarbonizing the EII, attention must also be given to the process emissions and the potential end-of-life emissions from products that contain carbon of fossil origin, such as petroleum-based plastics.

For the EII, three broad technical strategies for decarbonizing the production processes can be identified:

- *Biomass as fuel or feedstock:* Biofuels can replace fossil fuels in most processes and be used as feedstock for producing bio-based chemicals and materials, such as polymers. Biomass is readily available in the pulp and paper industry and has already replaced much oil use. If used in cement production, emissions can be reduced by about 50 per cent but the process emissions from calcium carbonate conversion remain. In principle, bio-coke can replace coal-based coke for reducing iron oxide to pig iron. But biomass and land is a limited resource and there are competing uses (for food, feed, fibre, chemicals and so on) and conflicts with other environmental objectives such as biodiversity and recreation. Bioenergy accounts for about 50 exajoules, or 10 per cent of current global primary energy use. The potential 2050 deployment levels have been estimated at 100 to 300 exajoules (IPCC, 2011) so the contribution compared to future global energy demand is limited.
- *Carbon Capture and Storage:* CCS for industrial application can reduce a large share of industrial emissions including process emissions. But applying CCS to industrial facilities, especially existing ones, is more complicated than applying CCS in the power sector. An industrial plant typically has several different source emissions with differing concentrations and the physical space for post-process capture CO_2-scrubbers may be limited. Currently proposed technologies do not capture all the CO_2 in the flue gases and they increase the consumption of heat and electricity. To capture more than about 80 per cent of all emissions from an industrial plant with CCS will require deeper integration into the core production processes. However, there are also some 'low hanging fruits' in terms of relatively pure CO_2-streams in some industrial processes. Many issues remain concerning CCS, including technical challenges, costs, large-scale infrastructure needs, legal aspects and lack of public acceptance (see Chapters 1 and 3).
- *Electrification:* Electrifying the process completely, or using hydrogen, is a radical solution that could rid the industry of fossil fuel-related emissions. A number of electro-thermal processes for industrial heating in different temperature ranges are possible (using, for example,

microwaves, infrared radiation or plasma). Hydrogen from electrolysis can be used for reducing iron oxide or replacing hydrogen from natural gas in fertilizer production. Through co-electrolysis of water and carbon dioxide, or by reacting hydrogen with carbon dioxide, a synthesis gas (mainly CO and H_2) can be produced from which a range of hydrocarbons and platform chemicals can be produced. Such 'power-to-gas', 'electro-fuels' or 'electro-plastics' processes are technically possible but relatively expensive. Industrial emission reductions from electrification rest on the assumption that electric power supplies are fully decarbonized (see Chapter 3).

In addition to these three basic options, conventional cement may, in principle, be replaced by alternative cements (for instance, by magnesium-based cement) or other building materials – eliminating the process emissions from calcination. For aluminium, research is ongoing, and has been for a long time for purely economic reasons, to develop inert anodes that would not cause process related CO_2 emissions from the depletion of the anodes, but the prospects for a breakthrough are uncertain.

On the issue of functional overlap we make the observation that CCS has no co-benefits in the EII since it is less expensive to release CO_2 to the atmosphere than to capture and transport it to final storage. Electrification and hydrogen may offer some process advantages but these are largely unexplored and they are likely to be trivial compared to the increase in energy cost relative to continued use of unabated fossil fuels. The bio-based industry is a more complex and interesting case. Decarbonization of transport will reduce demand for petroleum-based fuels and support the development of bio-refineries. At the same time, weak demand for some paper products is prompting the forest industry to look for new markets. Petroleum-based chemicals and materials will also be replaced eventually.

2.2 Co-evolution and interdependence with the energy system

The EII transforms and processes huge flows of materials that contain carbon (coal, oil, natural gas and biomass) for energy use as well as for use as raw material and feedstock. Changes in the use of energy and feedstock will have considerable implications for the rest of the energy system depending on which routes to decarbonization are chosen. Historically, the EII has co-evolved with the economy as a whole and with the development of energy systems. The European coal–steel nexus, the merger of the chemicals and petro-industry after World War I or the early uses of hydroelectricity to power new industries are some examples (Kaijser & Kander, 2013; Bennett & Pearson, 2009). For example, petroleum refineries in their present form are set to decrease production as an effect of climate policy. This will have implications for the chemical sector that relies on refinery by-products (for example, naphtha) as feedstock.

Sustainability concerns are already pushing the introduction of bio-based chemicals and fuels. Emerging bio-economy solutions for producing biofuels, chemicals and electricity could induce a stronger integration of sectors such as agriculture, forestry and chemicals but also increased competition for biomass. It is an open question whether the petrochemical, forestry, or some other industry will be the champion of the future bio-based economy.

Electrification of processes or a shift to hydrogen, for instance in steel and industrial heating, could lead to very large increases in total electricity demand. Such technology shifts will have profound energy system implications. Electrifying EU primary steel production alone would require an additional 380 TWh of electricity per year, equivalent to about 13 per cent of present total EU electricity demand. The total use of fossil fuels for 'non-energy consumption' which, as noted, amounts to 1190 TWh in energy terms needs eventually to be replaced.

3. The policy framework for governing industrial emissions

The climate policy debate has mainly been concerned with marginal emission reductions and costs and benefits in the near term (such as, the Kyoto Protocol or 2020 targets). The need to develop integrated policy strategies for a low-carbon transition in industry is a relatively new issue that has emerged as a result of the 2050 objective. Here, we describe and analyse the framework, as of 2014, for governing an industrial low-carbon transition. The key policy domains in this include climate and energy, trade and industrial policy.

3.1 Climate and energy policy

Industrial GHG emissions are mainly regulated via the EU Emissions Trading Scheme (ETS) that puts a cap on emissions within the EU. The Industrial Emissions Directive, IED (2010/75/EU) is a complementing regulation which is the basis for granting and updating industrial permits at the national level. The Industrial Emissions Directive takes an integrated approach (that is, the whole environmental performance of the plant), based on best available techniques, and allows flexibility to avoid subjecting plants to disproportionately high costs. In principle, the Industrial Emissions Directive could be used for regulating emissions at a lower total cost to industry (Johansson, 2006), but in practice the ETS is the main instrument for emission reductions in heavy industry.

The ETS includes all major industrial GHG-emitting facilities including heat and power production. Installations included in the EU ETS under the emission cap will, by definition, collectively reach the targeted reductions. If the EU is to meet its ambitions of 80 to 95 per cent reduction by 2050, the reduction rate of 1.74 per cent per year until 2020 needs to be increased. Discussions concerning new targets for 2030 that are aligned with the 2050 target are ongoing as of 2014, including an increase of the reduction rate to 2.2 per cent per year (European Commission, 2014a).

Under a tight and long-term emission cap, the real issue is not if the targets will be met, but how and at what cost. For industry, different outcomes are plausible and strongly dependent on complementary policies and other societal goals. Reducing emissions can be achieved by reducing output (and importing from abroad thus causing carbon leakage), by technical measures (often costly as discussed above) and through reductions in consumer demand for basic materials. The costs and future directions for mitigating GHG emissions in the industrial sector are also strongly influenced by energy policy at the EU and national levels.

The primary objective of the ETS is to reduce GHG emissions. However, a secondary objective is also, as a price-based mechanism, to promote improvements and thus innovation and adoption of new technologies. The price signal given by the ETS is important for providing economic incentives to mitigation measures but mainly through incremental technical changes. There is relatively broad consensus (see, for example, IPCC, 2014) that a carbon price alone is insufficient to support and induce the long-term technological shifts that are required. It needs to be complemented with policy instruments targeting technology development, demonstration and up-scaling (Hanemann, 2010).

In 2008, the EU launched the climate and energy package in an effort to integrate these two policy fields, expressed through the 20/20/20 targets (Chapter 1). From the perspective of industry, the effect of the 'energy component' of the package (including specific targets for energy efficiency and renewable energy) has been to reduce the mitigation costs by reducing the price of emission permits within the ETS. So far, the EII is largely exempt from the costs of support schemes for renewable energy sources (RES) and can usually benefit from energy efficiency policies (Ericsson et al., 2011; Stenqvist, 2014; see also Chapter 3). For decarbonization, even stronger climate and energy policy integration will be needed, with serious consideration of long-term climate objectives (Dupont & Oberthür, 2012).

3.2   Trade and industrial competitiveness

Trade policy must be coordinated with climate policy as long as there are no international climate policy agreements that provide a level playing field for industry (for example, in terms of universal carbon prices) (Helm et al., 2012). Progress on this issue is not likely in the near future since Article 3 of the United Nations Framework Convention on Climate Change lays down 'common but differentiated responsibilities' as an important principle. With differing national ambitions, industries that operate in global markets have to develop strategies spanning across countries with different requirements on emission reductions.

So far, the EU strategy to avoid carbon leakage has been to shelter the EII from policy-driven direct or indirect cost increases. The EII receives mainly free allocation of emission permits under the ETS and compensation for electricity price increases is allowed under the ETS Directive (2009/29/EC). Currently, 96 per cent of industrial GHG emissions within the EU ETS are on the 'carbon leakage list'.[3] Furthermore, the EII receives preferential tax treatment with lower energy taxes. Industry, as noted, is also often exempt from the burden of RES support schemes. Nevertheless, European industry is facing considerably higher energy prices than important competitors in countries such as the United States and China (IEA, 2013). The strategy to shield industry from the costs seems to have worked so far since there is no real evidence of carbon leakage from Europe in the first periods of trading up to 2012 (Bolsher et al., 2013). Special provisions will most likely remain in place to protect EII from additional costs that may disadvantage them in international markets (Neuhoff et al., 2014). However, future and more stringent climate policy will exacerbate the disadvantages, and the current compensation measures will be insufficient, as the carbon budget of the ETS gets tighter.

Energy price differences among countries and regions also arise due to differences in access to natural resources. The US unconventional gas boom, which led to much lower gas prices than in Europe, and reduced CO_2 emissions from US power production, is a case in point. Countries and regions endowed with RES may enjoy similar cost advantages and become home to low-emission energy-intensive manufacturing in the future.

Basic materials are traded internationally to varying degrees and they can be more or less sensitive to production cost increases. For example, cement and clinker, which is bulky and expensive to transport, is mostly produced and used locally and regionally. The annual export of cement from the EU in 2010–2012 was 11 to 17 Megatonnes (Mt), whereas the import was 3 to 6 Mt. The resulting average net export of about 10 Mt is relatively small compared to the annual EU production of about 200 Mt (EU Market Access Data Base). The annual export of paper 2010–2012 was 19 to 20 Mt and the import was 5 to 6 Mt. The resulting net export of about 15 Mt contrasts with a total production of about 90 Mt (CEPI, 2012). In addition to transport costs, production may be more or less tied to local raw materials (such as minerals or pulp wood), supporting infrastructure and human resources, and integrated and complex value chains – factors that make relocation difficult.

Although markets for many materials are relatively regional due to transport costs and other factors, this situation may change if ambitions to decarbonize the EII are followed through. In this scenario, solutions for maintaining production within the EU must be sought in the trade regime, including neighbourhood policy, and through industrial development policy. Border taxes may become inevitable unless various support schemes can be designed to compensate for climate policy related cost increases. Several of Europe's competitors subsidize capital and energy for the basic industry (Haley & Haley, 2013). Where non-EU countries give unfair support to their industries through tariff and non-tariff barriers, subsidies and export incentives, the EU can challenge such practices under the WTO. Where leakage is most likely to occur to neighbouring countries, like cement production in the MENA region, bilateral agreements through neighbourhood policy may be sufficient.

The relatively high mitigation costs for decarbonized basic materials production needs to be handled in the evolving global climate regime, and/or through border tax adjustments and compensation schemes within the EU to avoid carbon leakage. It must be noted that there is also the possibility of positive carbon leakage, in that production moves to 'low-carbon energy islands' with an abundance of RES. Iceland and the Nordic countries, with good wind resources and large hydropower or geothermal capacities, may be cases in point, as may be Brazil, with large hydropower, solar, wind and bioenergy resources.

So far, introducing trade-related compensation schemes for higher carbon cost in the EU, such as carbon border tax adjustments, have not been high on the agenda for the European Commission. A similar idea was introduced with the inclusion of international aviation in the ETS. This initiative stirred international controversy and was forcefully resisted by both the United States and China. European Commission actions in the trade arena linked to climate change have so far been limited to emerging technologies driven by climate policy, such as the conflict with China regarding unfair Chinese subsidies to solar photovoltaics (European Commission, 2014b).

Although there are many good reasons to avoid trade barriers, the conflict between imposing higher costs and maintaining free trade will continue to grow and will eventually force policy intervention. Some argue that carbon tariffs on imported goods from countries with weak climate policies could induce them to join international climate and trade agreements, and to implement climate policies (Helm et al., 2012). Decarbonizing industry will certainly put climate and trade policy to the test.

3.3   Industrial policy, innovation and technological development

Industrial policy has been around for a long time in Europe and elsewhere (Grabas & Nutzenadel, 2013). The concept is often associated with failed government attempts to rescue outdated or ailing industries, or to 'pick winners' in terms of technologies or emerging companies. Although industrial policy has a mixed record, industrial policy of some sorts is inevitable. Priorities and trade-offs in infrastructure investments, research and development, education, environmental protection, tax systems, financing, labour laws and other factors that determine the framework conditions or 'playing field' for industry need to be made. Within this broad definition of industrial policy, governments may choose to be passive or active but they cannot be neutral.

After the financial crisis in 2008, there was a renewed interest in industrial policy. In 2010, the European Commission tabled its 'Integrated Industrial Policy for the Globalisation Era, Putting Competitiveness and Sustainability at Centre Stage' (European Commission, 2010b), which 'sets out a strategy that aims to boost growth and jobs by maintaining and supporting a strong, diversified and competitive industrial base in Europe offering well-paid jobs while becoming more resource efficient'.[4] A distinction is often made between horizontal industrial policies that do not seek to promote any specific sector and vertical sector-specific policies. The EU industrial policy explicitly combines both by: 'bringing together a horizontal basis and sectoral application [...] and the Commission will continue to apply a tailor made approach to all sectors' (European Commission, 2010b, p. 4).

The EU integrated industrial policy has its roots in the Lisbon agenda (2000) – an attempt to make the EU the most competitive economy in the world – and it has co-evolved with broader objectives for sustainable development and concerns over reliable access to raw materials (European Commission, 2013d). It was followed in 2014 by a Communication that sets out the Commission's key priorities for industrial policy, 'For a European Industrial Renaissance' (European Commission, 2014c). This contributed to the European Council debate on industrial policy in June 2014 and contains a number of proposals on information, energy and transport infrastructure, the internal market, competitiveness proofing and regulatory fitness checks, innovation, finance, SMEs, international trade, standardization and so on. EU industrial policy, including raw materials, rarely mentions the EII and focuses mainly on emerging and down-stream sectors such as space, information and communications technology and rare-earth metals. The documentation of the high-level groups for iron and steel, chemicals and for raw materials does address the EII[5] with a focus on the challenges of increasing competition, the need for favourable conditions for these industries in the EU, and action against unfair practices in other countries.

Innovation and support for technical change along the whole innovation chain (including research & development (R&D), demonstration, pilot test and market formation support/early deployment) is also a key policy area linked to decarbonizing industry. The EU Framework Programmes for research have been replaced by a more holistic approach with the EU Horizon 2020 initiative. Horizon 2020 aims to integrate better the user side of research and support for market formation. The hope is that the EU can then more effectively translate academic research into usable products and services innovations. A number of programmes are relevant to the EII, such as the ultra-low carbon dioxide project for steel,[6] several efforts in the area of bio-economy, and the Strategic Energy Technology Plans for industry focusing on CCS and emerging down-stream technologies such as photovoltaics and wind, among others, but not yet for greening basic materials (see European Commission, 2013c). Financing mechanisms are also tied to some types of projects, for instance, via the European Investment Bank and the New Entrants Reserve programme (NER300). NER300 is an EU scheme to provide financial support to innovative renewable energy technology and CCS with the income from 300 million auctioned emission allowances under the ETS.

Current industrial R&D and innovation efforts both at the EU and member state level seem insufficient for the EII when considering the

2050 objective. One reason may be the conventional wisdom among innovation researchers that mature industries, usually engaged in incremental innovations, do not need public R&D support (Edquist & Chaminade, 2006). However, this 'wisdom' does not take into consideration the large-scale technology shifts needed to reach long-term climate objectives. Time scales, technology, political and economic risks are all well beyond previous experience when it comes to reducing GHG emissions to near zero by 2050. More long-term and sequential technology development and policy strategies are needed in order to develop the technologies for deep emission reductions beyond 2030. Present EU policy on technology development does not fully acknowledge the potential of non-CCS technologies for decarbonization. It becomes captured by incumbent interests (for example, that of energy companies in CCS) and their short-term priorities, and misses the long-term 2050 perspective.

4. Conditions for a policy-driven transition of the EII

The preceding analysis shows that the current combination of EU policies, overall, tends to preserve rather than create conditions for a low-carbon transformation of industry. Efforts to promote the bio-based economy may be an exception. As discussed above, the EII is generally sheltered from the effects of climate and energy policy and it is given various favourable exemptions to support competitiveness. Not much research, development and demonstration funding is going to the basic materials industry, and that which is does not appear to be directed at fundamental technology shifts in core processes.

The current situation can be understood partly through analysing the degree of functional overlap with other goals, or co-benefits, political will and societal interest. These interlinked and overlapping factors can provide part of the explanation of the current institutional set-up. Against this background, we discuss the prospects for a new institutional set-up in the next section.

Functional overlap is very important for gaining support and acceptance for policy. Its popularity is evident in its many labels: co-benefits, ancillary benefits, synergies, policy hitchhiking and piggybacking. The basic idea, from a decarbonization perspective, is that emission reductions generate other benefits and contribute to other environmental and societal goals such as air quality and job creation. Some measures with marginal emission reductions create clear synergies. For example, energy efficiency improvements, process or product changes, and increased materials efficiency can result in cost reductions, quality improvements and waste reduction.

Unfortunately, in the case of EII, functional overlaps hardly facilitate the zero-emissions target. Technology development for emission reductions may open up process advantages in some cases but it appears that decarbonization mainly inflicts substantially higher production costs. The degree of co-benefits is uncertain since options and strategies are relatively unexplored, but decarbonizing EII essentially means producing the same material or chemical compound in a more expensive way. CCS is a case in point: it brings no additional benefits, but a range of new worries. If future potential electricity and hydrogen use in industry can be flexible, it may be used to balance an electricity grid with more variable production. But we do not know to what extent future technologies and processes can handle variable power supply. The impact on jobs is very uncertain but likely to be small in this non-labour-intensive sector.

There is an overarching and relatively strong political will to decarbonize Europe, including industry. It is manifested through the ETS, which includes much of the basic industry, and in the 2020 and 2030 targets and the 2050 roadmaps. But the EU is far from united. Several member states have very ambitious long-term plans, such as the United Kingdom, Denmark and Germany, whereas others do not (see also Chapter 1). The policy implication of near-zero emissions is a quite recent issue, perhaps first brought to wider attention through the low-carbon economy roadmap (European Commission, 2011a). Governments have always been concerned about the risks of costs increases to industry from energy and climate policy, and taken measures to mitigate or compensate such effects. But completely decarbonizing industry makes such exemptions and compensation schemes much more difficult. We also note that there seems to be an emerging political will, a growing acceptance for, and change of narrative, when it comes to 'industrial policy' as indicated by the discourse on 're-industrialization' and 'green economy' in recent years.

Societal interest and involvement is also an important precondition for policy and governance. This is clear and present in areas such as food and climate, cars and urban transport, or lighting (consider, for example, the success of Earth Hour). These are areas that come very close to people and their everyday life. Two examples are environmental labelling of paper and the promotion of biodegradable plastic bags made from biopolymers. Whereas consumers have an interest in cars, they generally have no relation to the basic materials that cars are made of. Thus, societal interest in the basic materials industry is low and likely to remain so. Societal interest may be mirrored through NGOs but industry oriented NGO initiatives such as the WWF Climate Savers do not target or attract the EII. A rare exception to the generally low interest is environmental NGO Bellona that has taken an active stance pro-CCS, including industrial CCS.

Another aspect of societal interest, and part of the institutional set-up, is the role of labour unions. Basic industry has been integral to the development of the industrial state and job creation is an important social aspect. Labour unions organizing workers from the EII have typically held a strong political voice. Today, the divide in industry between the down-stream manufacturing industry that will most likely benefit from climate policy and the up-stream EII that faces a tough time is reflected in the various positions and attitudes towards climate policy in different labour unions. For example, IG Metal in Germany is quite progressive and sees opportunities in a transition whereas IG BCE (which stands for Bergbau, Chemie, Energie), which includes coal industry interests, is resisting change.

In summary, the current framework for governing emissions in industry rewards incremental improvements rather than prepares for the transition needed for decarbonization. This is partly because near-zero emissions by 2050 is a new idea. Protecting competitiveness and jobs is also an important and legitimate explanation, and there are good reasons for keeping basic materials production in Europe. With carbon leakage, nothing is gained. Integration along value chains is an important source of innovation and in some cases it is in the EU that we find the feedstock (such as metal ores or wood, but also potentially clean energy for the processes). Maintaining production capacity for certain metals, fertilizers and so forth, also has a supply security argument.

5. Towards a new institutional set-up

An important first step in a transition process is to develop some sort of shared vision, scenarios and clear direction for the longer-term development. Such visions are relatively well established in areas such as energy supply, smart grids, transport and buildings, although some of the details may be disputed. For basic industries this process has just started through the European Commission roadmap and subsequent industry subsector roadmaps.

Overcoming the barriers to low-carbon technologies in basic industry and at the same time managing the risk of carbon leakage requires a comprehensive and systemic policy approach. It includes the development of new EU internal and external policy strategies that inte-grate industrial, technology development, trade and climate policy. The need for such new policy strategies is articulated by the Confederation of European Paper Industries (CEPI, 2011, p. 3):

> A new level of climate policies is needed: to achieve the reduction required while avoiding carbon leakage, policies need to be har-monised with global developments and industry investments cycles. The EU needs to complement the current carbon price and target-based policy approach with a multi-dimensional and industry specific climate change policy. The policy package should include a tech-nology focus, be synchronized with industry investment cycles and global action, and include a raw material and product perspective.

Research, development and demonstration and up-scaling for technol-ogy development and deployment require large investments. A major obstacle is lack of financing for up-scaling and moving breakthrough technologies beyond the demonstration-phase. Another is the lack of regulatory frameworks that reduce investment risks through creating a trustworthy future market environment, as feed-in-tariffs have pro-vided for renewable electricity (see, for example, Burnham et al., 2013). The risk of deploying new technologies and processes is thus not only technological but also political. Will climate policy persist and is there a trustworthy regulatory and market environment? The importance of this is illustrated by experiences from the NER300 scheme where a number of granted projects have been cancelled due to the lack of clear direction in future markets for renewable energy (such as biofuels for transport), or for carbon emissions and CCS. A similar scheme to the NER300 could be used for financing future technology demonstra-tions in the EII but it would only be one piece of an integrated policy package.

Policies and investment strategies in the capital intensive EII also need to consider the large sunk cost in existing facilities and the complexity of operations. Core industrial processes change only gradually over the years and the investment cycles in heavy industry are long. For many of these companies, 2050 is only one or two major investment deci-sions away. EII must aim at becoming 'zero emissions ready' by 2020 or 2030, meaning that technologies have been developed and proven. Major investments in new core processes can be made thereafter, start-ing perhaps around 2030, provided that the broader institutional and market conditions make them economically viable.

Deployment through major investments is dependent on market conditions, which in turn are contingent on EU and international climate, energy and trade policies. It requires a coordinated response where climate policy and industrial policy is well integrated also into the EU response for maintaining open trade on a fair basis. This includes the use of bilateral agreements and various trade arrangements for easing the risk of unacceptable carbon leakage (some carbon leakage may be acceptable, and relocation of industry to regions with low-carbon energy supply may be welcomed).

The new global climate policy framework that is expected to emerge in 2015 may have implications for EU strategies for the EII. We do not want to speculate about the outcome of the international negotiations but it is clear that issues concerning carbon leakage and the EII must be dealt with if the EU wants to maintain its 2050 goals. Decarbonizing and keeping industry in Europe can be seen as in line with sustainable development, since the alternatives are clearly unsustainable. EU investments made in developing low-carbon process technologies could later benefit other countries, analogous to the development of renewable energy technologies, and be seen as a major contribution to international climate protection.

Conclusions

Decarbonization provides opportunities for much of industry to innovate and adapt to new and 'green' market demands. For energy intensive industries, however, decarbonization requires innovation and new investments in core process technologies that offer few co-benefits. There is not yet a shared vision and clear direction for a low-carbon transition in the energy intensive industries. There are gaps in key steps of the innovation chain, including insufficient R&D on basic options, such as electro-thermal technologies, lack of financing mechanisms (although NER300 may be further developed and expanded) and a need to create stable market conditions for green but more expensive basic materials. Although the ETS provides a basis for nudging industry towards lower emissions, it falls short of inducing the required longer-term technology shifts. It simply is not geared towards supporting a low-carbon transition in industry. Compared to other sectors, the EII faces greater economic, policy and governance challenges. The lack of co-benefits and societal interest, conflicts with free trade ideals, historical experience with industrial policy and hesitance concerning the role of government in industrial restructuring are factors that impede the development of a new approach. We have discussed some of the key aspects of such an approach in this chapter. In essence, it requires the development of a new EU-internal policy that integrates innovation, industrial, and climate and energy policy for decarbonization. It also requires an EU external policy that integrates international trade, foreign (for instance, neighbourhood policy and development cooperation), climate and energy policy.

Notes

1. This work was supported by the Swedish Energy Agency through the programme on International Climate Policy (project on Green Transition and Co-Evolution of Industry and the Energy System) and the Research Council of Norway through collaboration with CICEP (Strategic Challenges in International Climate and Energy Policy). We would like to thank the editors and Stefan Lechtenböhmer for valuable comments and discussions.
2. Back-stop technology is often used to denote a technology to fall back on, with quite unlimited potential but relatively expensive, if other less costly options fail or have been exhausted. Although CCS is sometimes referred to as a back-stop technology, it should be noted that storage sites are not unlimited.
3. The carbon leakage list includes firms under the ETS that will receive free allocation of emission permits up to the benchmark.
4. See http://ec.europa.eu/enterprise/policies/industrial-competitiveness/industrial-policy/index_en.htm, date accessed 10 February 2014.
5. There are several high-level groups between the European Commission and stakeholders, including one for steel, one for chemicals (concluded 2009), one on industrial policy and competitiveness and a High Level Steering Group on innovation and raw materials. These groups are formed ad hoc and serve as forums for information exchange between the Commission and industry.
6. ULCOS, Ultra-Low Carbon dioxide (CO_2) Steelmaking, is a joint EU-industry research programme. See www.ulcos.org, date accessed 14 October 2014.

References

Åhman, M., Nikoleris, A. & Nilsson, L. J. (2012) 'Decarbonising Industry in Sweden – An Assessment of Possibilities and Policy Needs', *EESS report*, No. 77 (Lund: Lund University).

Bennett, S. J. & Pearson, P. J. G. (2009) 'From Petrochemical Complexes to Biorefineries? The Past and Prospective Co-evolution of Liquid Fuels and Chemicals Production in the UK', *Chemical Engineering Research and Design (ChERD)*, *87*(9), 1120–1139.

Bergman M., Schmitz A., Hayden M. & Kosonen K. (2007) 'Imposing a Unilateral Carbon Constraint on Energy-Intensive Industries and Its Impact on their International Competitiveness – Data and Analysis', *European Economy Economic Papers, Number 298* (Brussels: DG-Economic and Financial Affairs).

Bolsher, H., Graichen, V., Graham, H., Healy, S., Lenstra, J., Meindert, L., Regerczi, D., v. Schickfus, M., Schuacher, K. & Timmons-Smakman, F. (2013)

'Carbon Leakage Evidence Project – Fact Sheet for Selected Sectors', (Rotterdam: ECORYS) http://ec.europa.eu/clima/policies/ets/cap/leakage/docs/cl_evidence_factsheets_en.pdf, date accessed 8 October 2014.

Boschma, R. (2005) 'Proximity and Innovation: A Critical Assessment', *Regional Studies, 39*(1), 61–74.

Burnham, J., Debande, O., Jones, O., Mihai, C., Moore, J. & Temperton, I. (2013) 'Report on Innovative Financial Instruments for the Implementation of the SET Plan, First-of-a-kind project', *JRC Scientific and Policy Reports*, Ref: EUR 26058, OPOCE LD-NA-26058-EN-N, https://ec.europa.eu/jrc/sites/default/files/ldna26058enn_002.pdf, date accessed 8 October 2014.

CEFIC (2013) *European Chemistry for Growth: Unlocking a Competitive, Low Carbon and Energy Efficient Future* (Brussels: CEFIC), http://www.cefic.org/Documents/PolicyCentre/Energy-Roadmap-The%20Report-European-chemistry-for-growth.pdf, date accessed 8 October 2014.

CEMBUREAU (2013) *The Role of Cement in the 2050 Low Carbon Economy* (Brussels: CEMBUREAU), http://www.cement.ie/brochure/cembureau-brochure.pdf, date accessed 8 October 2014.

CEPI (2011) *Unfold the Future. The Forest Fibre Industry – 2050 Roadmap to a Low-Carbon Bio-Economy* (Brussels: Confederation of European Paper Industries), http://www.unfoldthefuture.eu/uploads/CEPI-2050-Roadmap-to-a-low-carbon-bio-economy.pdf, date accessed 8 October 2014.

CEPI (2012) *CEPI Key Statistics 2012*, http://www.cepi.org/topics/statistics, date accessed 8 October 2014.

Dupont, C. & Oberthür, S. (2012) 'Insufficient Climate Policy Integration in EU Energy Policy: The Importance of the Long-term Perspective', *Journal of Contemporary European Research, 8*(2), 228–247.

EAA (2012) *An Aluminium 2050 Roadmap to a Low-Carbon Europe: Lightening the Load* (Brussels: European Aluminium Association), http://www.alufenster.at/rte/upload/wohnbau_neu/an-aluminium-2050-roadmap-to-a-low-carbon-europe.pdf, date accessed 8 October 2014.

Edquist, C. & Chaminade, C. (2006) 'Industrial Policy from a Systems-of-innovation Perspective', *EIB Papers: An Industrial Policy for Europe? Context and Concepts, 11*(1), 108–132.

Ericsson, K., Nilsson, L. J. & Nilsson, M. (2011) 'New Energy Strategies in the Swedish Pulp and Paper Industry', *Energy Policy, 39*, 1439–1449.

Eurofer (2013) *A Steel Road Map for a Low Carbon Europe 2050* (Brussels: Eurofer).

European Commission (2010b) 'An Integrated Industrial Policy for the Globalisation Era: Putting Competitiveness and Sustainability at Centre Stage', COM(2010) 614.

European Commission (2011a) 'A Roadmap for Moving to a Competitive Low Carbon Economy in 2050', COM(2011) 112.

European Commission (2011b) 'Energy Roadmap 2050', COM(2011) 885.

European Commission (2013a) *Competing in Global Value Chains – EU Industrial Structure Report 2013* (Luxembourg: Publications Office of the European Union).

European Commission (2013b) 'Sixth National Communication and First Biennial Report from the European Union under the UN Framework Convention on Climate Change (UNFCCC)', Technical Report 2014–075, C(2014) 3.

European Commission (2013c) 'European Industrial Initiatives and Innovation', http://ec.europa.eu/energy/technology/initiatives/initiatives_en.htm, date accessed 8 November 2014.

European Commission (2013d) 'On the Implementation of the Raw Materials Initiative', COM(2013) 442.

European Commission (2014a) 'A Policy Framework for Climate and Energy in the Period from 2020 to 2030', COM(2014) 15.

European Commission (2014b) 'Press Release: Commissioner De Gucht: "We Found an Amicable Solution in the EU-China Solar Panels Case That Will Lead to a New Market Equilibrium at Sustainable Prices"', MEMO/13/729, 26/07/2013.

European Commission (2014c) 'For a European Industrial Renaissance' COM(2014) 14/2.

Glass for Europe (2013) 'Europe's Flat Glass Industry in a Competitive Low Carbon Economy: Performance, Sustainability, Capacity to Help Deliver Europe's Low Carbon Future', http://www.glassforeurope.com/images/cont/214_51388_file.pdf, date accessed 8 October 2014.

Grabas, C. & Nutzenadel, A. (2013) 'Industrial Policies in Europe in Historical Perspective' *Working Paper No. 15* (Brussels: DG Research and Innovation) http://www.wifo.ac.at/jart/prj3/wifo/resources/person_dokument/person_dokument.jart?publikationsid=46867&mime_type=application/pdf, date accessed 8 October 2014.

Haley, U. & Haley, G. (2013) *Subsidies to Chinese Industry: State Capitalism, Business Strategy and Trade Policy* (Oxford: Oxford University Press).

Hanemann, M. (2010) 'Cap-and-trade: A Sufficient or Necessary Condition for Emission Reduction?' *Oxford Review of Economic Policy, 26*(2), 225–252.

Hansen, T. & Winther L. (2011) 'Innovation, Regional Development and Relations between High- and Low-tech Industries', *European Urban and Regional Studies, 18*, 321–329.

Helm, D., Hepburn, C. & Ruta, G. (2012) 'Trade, Climate Change, and the Political Game Theory of Border Carbon Adjustments', *Oxford Review of Economic Policy, 28*(2), 368–394.

IEA (2013) *World Energy Outlook* (Paris: International Energy Agency/OECD).

IPCC (2011) *Special Report on Renewable Energy for Climate Change Mitigation* (Geneva: Intergovernmental Panel on Climate Change), https://www.ipcc.ch/pdf/special-reports/srren/SRREN_FD_SPM_final.pdf, date accessed 8 October 2014.

IPCC (2014) *Climate Change 2014: Mitigation of Climate Change. Contribution of Working Group III to the Fifth Assessment Report of the Intergovernmental Panel on Climate Change* (Cambridge and New York: Cambridge University Press).

Johansson, B. (2006) 'Climate Policy Instruments and Industry – Effects and Potential Responses in the Swedish Context', *Energy Policy, 34*(5), 2344–2360.

Kaijser, A. & Kander, A. (2013) 'Framtida Energiomställningar i ett Historiskt Perspektiv' (Future Energy Transitions in an Historical Perspective), *Rapport 6550* (Stockholm: Naturvårdsverket).

Lapillone, B., Pollier, K. & Sebi, K. (2012) 'Energy Efficiency Trends in Industry in the EU – Lessons from the ODYSSEE MURE Project', http://www.odyssee-mure.eu/publications/br/energy-efficiency-trends-industry.html, date accessed 8 October 2014.

Lempert, R. J., Popper, S. W., Resetar, S. & Hart, S. (2002) 'Capital Cycles and the Timing of Climate Change Policy', *PEW Centre on Global Climate Change,* https://www.greenbiz.com/sites/default/files/document/CustomO16F37186. pdf, date accessed 8 October 2014.

Neuhoff, K., Acworth, W., Dechezleprêtre, A., Dröge, S., Sartor, O., Sato, M., Schleicher, S. & Schopp, A. (2014) 'Staying with the Leaders – Europe's Path to a Successful Low-carbon Economy', *Climate Strategies,* http://www.swp-berlin. org/fileadmin/contents/products/fachpublikationen/Droege_staying_with_ the_leaders_AcrobatNochmal2.pdf, date accessed 8 October 2014.

Nilsson, M., Nilsson, L. J., Hildingsson, R., Stripple, J. & Eikeland, P. O. (2011) 'The Missing Link: Bringing Institutions and Politics into Energy Future Studies', *Futures, 43*(10), 1117–1128.

Stenqvist C. (2014) *Industrial Energy Efficiency Improvement: The Role of Policy Evaluation,* (PhD Thesis) (Lund: Lund University).

6
Transport: Addicted to Oil

Tom van Lier and Cathy Macharis

Introduction

Transport fulfils a crucial role for both economic and social development, since it enables people and goods to move from one place to another. Globalization and technological innovation have led to a drastic increase in national and international freight and passenger transport and consequently also in transport-related greenhouse gas (GHG) emissions. In this chapter, we first discuss the historical developments in transport in the EU, the sector's potential role in decarbonization and the current EU policies with regard to decarbonization in the transport sector. Second, we identify a number of policy gaps, since current policies are most likely insufficient to achieve the targeted decarbonization in the transport sector by 2050. Third, we analyse the drivers and barriers to achieving the transport-sector reduction targets focusing on functional overlap, political will, societal backing and institutional set-up.

1. Historical development

Between 1995 and 2011, passenger transport in the European Union (EU-27) – excluding transport activities between the EU and the rest of the world – increased by 22.5 per cent to 6.57 billion passenger kilometres, which is equivalent to, on average, 13,060 km per person. Of the total, passenger cars accounted for 73.4 per cent; buses and coaches 7.9 per cent; railways 6.3 per cent; powered two-wheelers 1.9 per cent; and tram and metro 1.4 per cent. Intra-EU air and intra-EU maritime transport contributed 8.8 per cent and 0.6 per cent, respectively (European Commission, 2013, p. 19). The average European travelled almost 9,500 km by car in 2010 (European Commission, 2013, p. 109). In 2011, 950 billion EUR, or roughly 13 per cent of the total household consumption, was spent on transport-related items (European Commission, 2013, p. 19). Freight transport experienced similar growth. In 2011, the total transport of goods in the EU-27 amounted to 3,824 billion tonkilometres. Road transport is also the dominant transport mode for freight, accounting for 45.3 per cent of the total, followed by intra-EU maritime transport (36.8 per cent). Rail accounted for 11 per cent, inland waterways for 3.7 per cent, oil pipelines for 3.1 per cent and intra-EU air transport accounted for 0.1 per cent (European Commission, 2013, p. 19).

Besides personal mobility and economic growth, transport also generates important negative side effects such as air pollution, noise nuisance, traffic accidents, congestion and GHG emissions. Transport accounted for 1,215.6 million tonnes of carbon dioxide equivalent (CO_2eq) out of a total of 5,005.8 million tonnes CO_2eq in 2010, or around a quarter of EU-27 GHG emissions, making it the second biggest GHG-emitting sector after energy. Road transport contributed almost one-fifth of the EU's total GHG emissions and more than two-thirds (72.1 per cent) of transport-related GHG emissions (European Commission, 2013, p. 122). Emissions from maritime navigation accounted for 14.1 and 12.4 per cent of transport-related GHG emissions, respectively. Compared to 1990 levels, GHG emissions in navigation and aviation have grown fastest. GHG emissions from road transport have also increased. Only emissions from rail transport and from other smaller transport sectors including pipeline transportation, ground activities in airports and harbours, and off-road activities have shown a declining trend (European Commission, 2013, p. 122). Figure 6.1 shows the relative share of transport and its components in the EU-27 GHG emissions in 2009.

The lack of progress in reducing GHG emissions in transport indicates that decarbonizing the sector constitutes a major challenge: while GHG emissions from other sectors decreased on average 24 per cent between 1990 and 2009, emissions from transport increased by 29 per cent in the same period, the highest percentage increase of all energy-related sectors (Hill et al., 2012).

To tackle the increasing GHG emissions in transport, it is important to understand the drivers of the sector. Since moving passengers or goods around serves wider social and economic objectives, transport demand is driven by a range of external factors. For example, GDP growth and increasing personal incomes, globalization, tourism, urbanization, population growth, employment rates, ICT evolution, decreasing real cost

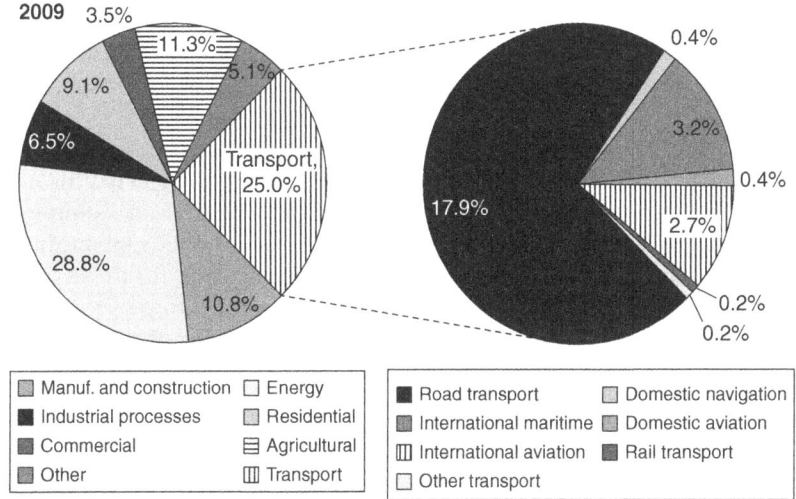

Figure 6.1 EU-27 greenhouse gas emissions by sector and mode of transport, 2009

Note: International aviation and maritime shipping only include emissions from bunker fuels.

Source: Hill et al. (2012) based on data from the European Environment Agency (EEA) GHG data viewer.

of transport and increasing speed of transport (Skinner et al., 2010) are generally considered to lead to increases in transport demand (although a decoupling may be possible in principle). Some external drivers such as higher energy prices can potentially reduce demand for transport, but the long-term trend in energy prices, in real terms, has been declining. Other factors might either increase or reduce transport demand. New infrastructure can lower GHG emissions if it encourages the use of less GHG intensive modes but can also increase GHG emissions if it increases overall capacity or reduces travel times (Skinner et al., 2010).

1.1 European policy vision for transport

While significant global action on transport does not exist, the EU has wide competence on transport and climate change and tends to act when there is a transnational element involved. The Treaty on the Functioning of the European Union (TFEU) explicitly states that competence for transport and climate change is shared between the EU and its member states. The EU's objectives in transport are pursued within the framework of a common transport policy and the TFEU sets out the legal basis for the exercise of competence applying to all areas of land transportation. Below, we discuss the European policy vision documents aimed at directly or indirectly tackling GHG emissions from transport as of 2014.

The 2020 climate and energy package, which aims to ensure the EU meets its climate and energy targets for 2020 (see Chapter 1) has incorporated elements specifically targeting transport (European Commission, 2008). The Effort Sharing Decision (No 406/2009/EC) establishes binding annual GHG emission targets for member states for the period 2013–2020. These reduction targets concern emissions from sectors not included in the EU Emissions Trading System, such as transport (except aviation and international maritime shipping), buildings, agriculture and waste. Under the Renewable Energy Directive (2009/28/EC), Member States have taken on binding national targets for raising the share of renewable energy in their energy consumption by 2020, including a 10 per cent share of renewable energy in the transport sector.

In January 2014, the European Commission presented its proposal for a 2030 framework for climate and energy policies (European Commission, 2014). It states that emissions from sectors outside the EU ETS, including transport, would need to be cut by 30 per cent below the 2005 level in 2030, shared equitably among member states. Transport is recognized in the 2030 framework as a key complementary policy:

> Further reduction of emissions from transport will require a gradual transformation of the entire transport system towards a better integration between modes, greater exploitation of the non-road alternatives, improved management of traffic flows through intelligent transport systems, and extensive innovation in and deployment of new propulsion and navigation technologies and alternative fuels. This will need to be supported by a modern and coherent infrastructure design and smarter pricing of infrastructure usage. Member States should also consider how fuel and vehicle taxation can be used to support greenhouse gas reductions in the transport sector in line with the Commission's proposal on the taxation of energy products.
>
> (European Commission, 2014, p. 14)

The 2030 framework takes into account the longer term perspective set out by the Commission in the Roadmap for moving to a competitive low carbon economy in 2050 (Low-Carbon Roadmap; European Commission, 2011b), the Energy Roadmap 2050 (European

Commission, 2011c) and the Transport White Paper (European Commission, 2011d). These documents reflect the EU's goal of reducing GHG emissions by 80–95 per cent below 1990 levels by 2050 (see Chapter 1). For the transport sector (including aviation but excluding maritime shipping), Roadmap targets for 2030 GHG emissions are between plus 20 per cent and minus 9 per cent, and for 2050 between minus 54 per cent and minus 67 per cent compared to 1990 levels (while 2005 levels for transport were considered to be 30 per cent above 1990 levels) (European Commission, 2011b, p. 6).[1]

The Low-Carbon Roadmap identified improved fuel efficiency as the most likely main driver for reversing the trend of increasing GHG emissions in the transport sector to 2025. To this end, the amended Fuel Quality Directive (Directive 2009/30/EC) requires that well-to-wheel GHG emissions per unit of energy supplied be reduced by at least 6 per cent, and up to 10 per cent, by 2020. In order to bring emissions from road, rail and inland waterways to below 1990 levels in 2030, increased fuel efficiency should be combined with measures such as pricing schemes to tackle congestion and air pollution, infrastructure charging, intelligent city planning and improving public transport. In addition, improved efficiency and better demand-side management, fostered through CO_2 standards and smart taxation systems, are required to advance the development of hybrid engine technologies and facilitate the gradual transition towards cleaner vehicles in all transport modes, including plug-in hybrids and electric vehicles (powered by batteries or fuel cells) at a later stage. Biofuels are regarded as an alternative fuel especially in aviation and heavy-duty trucks, with expected strong growth in these sectors after 2030 (European Commission, 2011b).[2]

In the Energy Roadmap 2050 the Commission explores the challenges posed by delivering the EU's decarbonization objective while ensuring security of energy supply and competitiveness (European Commission, 2011c). With regard to transport, this roadmap stresses the importance of energy efficient vehicles and incentives for behavioural change, and points out that electricity will have to play a much greater role in the future.

The Transport White Paper presents the European Commission's vision for the future of the EU transport system and defines a policy agenda for the next decade to increase the competitiveness of transport within the EU while simultaneously moving towards a 60 per cent reduction in CO_2 emissions and a comparable reduction in oil dependency by 2050 in the transport sector (European Commission, 2011d).

The essence of the plan is to change the oil dependency of the transport system without sacrificing its efficiency or endangering mobility. Emission reduction objectives are set at 50 per cent for aviation, 40–50 per cent for navigation and 70–80 per cent for road transport, while 50 per cent growth is expected in passenger transport and 80 per cent growth in freight transport. The focus is on multimodality and creating conditions for modal shift by means of a fully integrated transport network.

The White Paper sets ten benchmark goals, underpinned by 40 concrete initiatives, for achieving a competitive and resource efficient transport sector. It divides the transport market into three important sections: medium distances, long distances and urban transport. The key points include developing and deploying new and sustainable fuels and propulsion systems; optimizing the performance of multimodal logistic chains (including making greater use of more energy efficient modes); and increasing the efficiency of transport and of infrastructure use through information systems and market-based incentives. The various actions and measures introduced within the Paper need to be elaborated by 2020 through the preparation of appropriate legislative proposals (European Commission, 2011d).

In the Roadmap to a Resource Efficient Europe (European Commission, 2011e), the Commission states that initiatives in the Transport White Paper need to be implemented consistently with resource efficiency objectives, particularly by moving towards internalization of external costs. This roadmap is one of the main building blocks of the resource efficiency flagship initiative, which in turn is part of the Europe 2020 Strategy for smart, sustainable and inclusive growth (European Commission, 2011a).

1.2   EU policies to lower emissions from transport

The EU has (as of 2014) a range of policies in place to lower emissions from the transport sector:

- Inclusion of aviation in the EU Emissions Trading System (ETS) (Directive 2008/101/EC);
- Passenger car CO_2 Regulation (Regulation (EC) No 443/2009 and Regulation (EU) No 333/2014);
- Light duty vehicles (vans) CO_2 Regulation (Regulation (EU) No 253/2014 and Regulation (EU) 510/2011);
- Clean road vehicles Directive (Directive 2009/33/EC);
- Renewable Energy Directive (Directive 2009/28/EC);

- GHG intensity reduction requirement of amended Fuel Quality Directive (Directive 2009/30/EC);
- Road Infrastructure Charging – Heavy Goods Vehicles (Directive 1999/62/EC as modified by Directive 2006/38/EC and by Directive 2011/76/EU);
- Introduction of rolling resistance limits and tyre labelling and mandatory installation of tyre pressure monitors on new vehicles (EU Tyre Labelling Regulation 1222/2009);
- Vehicle procurement rules for public authorities taking into account lifetime energy use and CO_2 emissions (European Commission, 2011f).

Additionally, the European Commission is working on a comprehensive strategy to reduce CO_2 emissions from heavy goods vehicles in both freight and passenger transport.

The regulatory framework for road infrastructure charging sets common rules on distance-related tolls and time-based user charges (vignettes) for heavy goods vehicles (above 3.5 tons) for the use of certain infrastructure. These rules stipulate that the cost of constructing, operating and developing infrastructure can be leveraged through tolls and vignettes to road users. Tolls may also include an external cost charge that reflects the cost of air and noise pollution (within certain limits). Although not directly focusing on GHG emission reduction, this framework might reduce vehicle kilometres and thus related fuel consumption.

Some additional policy components exist that might also influence GHG emissions in the transport sector. The development of a Trans-European Network in the Transport sector originates from the beginning of the 1990s, when the then 12 member states decided to set up an infrastructure policy at Community level in order to support the functioning of the internal market through continuous and efficient networks in the fields of transport, energy and telecommunications. Since January 2014, the EU has a new transport infrastructure policy that aims to close the gaps among member states' transport networks, remove bottlenecks to the smooth functioning of the internal market, and overcome technical barriers such as incompatible standards for railway traffic.[3] This new Trans-European Transport Network policy sets out a core network corridor approach (propagated in the Transport White Paper) that is also explicitly linked to the EU long-term transport policy objectives of meeting future mobility needs while ensuring resource efficiency and reducing carbon emissions.

Another important element stems from the taxation of energy products. On 13 April 2011 the European Commission, to promote energy efficiency and consumption of more environmentally friendly products, presented a new proposal for the taxation of energy products and electricity in the EU that should replace the 2003 EU Energy Taxation Directive (Directive 2003/96/EC). The proposed rules aim to restructure the way energy products, including motor fuels, are taxed to take into account both CO_2 emissions and energy content and to remove current imbalances that distort the Single Market.

2. Policy gaps

The EU policies to lower emissions from transport in place in 2014 as described above are not part of an overarching strategy or goal and are in themselves insufficient to reach the EU GHG emission reduction targets defined in the Low-Carbon Roadmap 2050 and the Transport White Paper. It is clear that more significant reductions in GHG emissions from transport are required.

A project undertaken for DG Climate Action investigated the policies and technologies needed to achieve substantial emission reduction by 2050.[4] To know the magnitude of the required effort, it projected the growth in transport's life cycle GHG emissions by mode according to a business as usual (BAU) scenario. While, in 2010, continued growth in the EU-27's transport-related GHG emissions was anticipated (Skinner et al., 2010), the updated 2011 baseline anticipated a decline in GHG emissions to 2050, mainly because of a range of existing and planned policies, including the 2020 regulatory CO_2 emissions targets for passenger cars and vans, the improvement of targets for maritime shipping based on the International Maritime Organization Energy Efficiency Design Index and estimated impacts of including aviation in the EU ETS (Hill et al., 2012). Road freight was now projected to slightly decrease by 2050 due to significantly reduced levels of demand growth and some additional modal shift. Even taking into account recent policy developments, emissions by aviation and shipping were expected to increase by 42 per cent and 22 per cent respectively between 2010 and 2050 without additional policy instruments (Hill et al., 2012).

Overall, the updated 2011 baseline shows a decline of 22 per cent in GHG emissions over the period to 2050 compared to 2010 GHG levels, but even in this more optimistic scenario, transport's well-to-wheel GHG emissions would still be 17 per cent higher in 2050 than in 1990 (Hill et al., 2012). This implies that additional action is required in order for the EU to meet its long-term GHG emission reduction targets in 2030 and 2050.

Hill et al. (2012) also demonstrated that transport emissions could reach levels around 20 per cent of economy-wide 1990 GHG emissions by 2050, if no additional action is taken, which would be equivalent to the total EU-wide GHG emissions budget for an 80 per cent reduction target across all sectors (Figure 6.2). The envisaged emission targets for transport of overall around 60 per cent are indicated in the lower part of Figure 6.2. Although it assumes simplifying linear reductions, Figure 6.2 clearly demonstrates the need for additional policy instruments in order to realize GHG emission reduction targets for the transport sector. Other studies confirm this need for additional efforts (Rijkee & van Essen, 2010; Geurs et al., 2011).

Skinner et al. (2010) estimated the potential GHG emission reductions that could be achieved in the transport sector through additional policy efforts. In a scenario where conventional fuels are substituted by biofuels (under the assumption that biofuels could achieve well-to-wheel average GHG emission savings of 85 per cent by 2050 and taking into account the maximum production potential estimated for the EU), transport GHG emissions are still expected to be almost 30 per cent higher in 2050 than in 1990, but slightly lower than transport's GHG emissions in 2010 (and 26 per cent lower than in a BAU scenario). The GHG savings potential for any biofuel is, however, very sensitive to the feedstock used and the way in which the biofuel is produced, as well as the method used to calculate the savings.

In another scenario, improvements in the technical energy efficiency of vehicles could deliver a reduction of 12 per cent in transport GHG emissions of 1990 levels by 2050 (50 per cent compared to BAU). This scenario assumes the virtual elimination of pure internal combustion engines from the vehicle fleet by replacing them with hybrids, plug-in hybrids, electric and fuel cell cars. For other modes, shifts to alternative energy carriers such as electricity and hydrogen were assumed where considered possible. In this scenario, the production of electricity and hydrogen for transport were considered to be essentially carbon-neutral (see Chapter 3).

A scenario that assumes the implementation of all technical options, that is to say, substituting conventional fuels with biofuels and achieving very significant improvements of vehicles' technical energy efficiency, would allow GHG emission savings of 36 per cent compared to 1990 by 2050 (63 per cent compared to BAU). The reduction

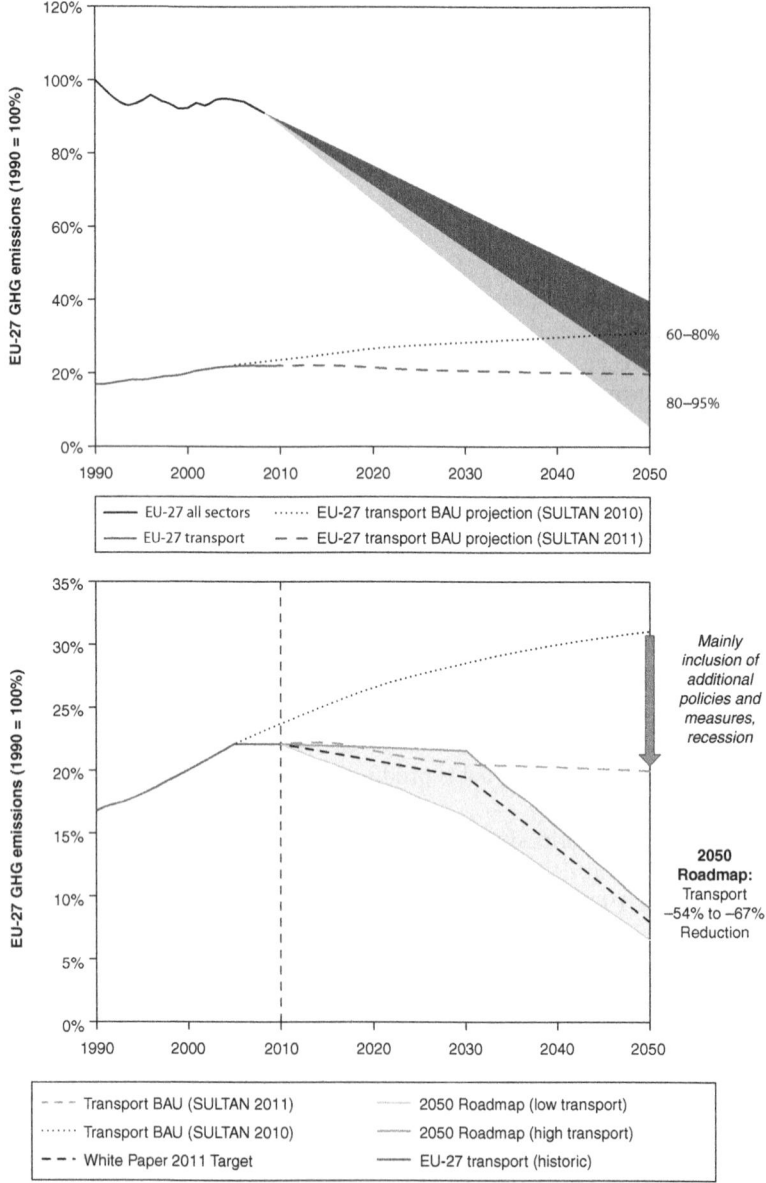

Figure 6.2 EU overall emissions trajectories against transport emissions (indexed)

Source: Hill et al. (2012) (based on EEA and SULTAN Illustrative Scenarios Tool, available at www.eea.europa.eu and www.eutransportghg2050.eu, date accessed 8 July 2014).

potentials for biofuels and improvements in the technical efficiency of vehicles are not additional, as the uptake of one option reduces the impact of the other. Also, because of interactions between individual measures, the order in which they are applied will affect their relative effectiveness (Skinner et al., 2010).

Achieving a reduction in transport GHG emissions by 50 per cent or more through the uptake of technical options alone is very difficult. Only in scenarios where both non-technical and technical options were taken up, GHG emissions from transport could be reduced by around 89 per cent by 2050 compared to 1990 (94 per cent on BAU). Non-technical options and relevant behavioural changes include improving the efficiency of vehicle use (optimizing speeds and routes, eco-driving, optimizing vehicle utilization), using the most appropriate mode for each (part of the) journey (co-modality) and increasing the efficiency of the transport system as a whole (structure and planning of transport system, demand restrictions, etc.). Overall, three broad approaches are simultaneously required to reduce transport's GHG emissions: reducing GHG intensity of energy used by the transport sector, improving the efficiency of transport vehicles by both technical and operational means, and improving the efficiency of the transport system (Skinner et al., 2010).

3. Analysis

In this section, we discuss – in accordance with the analytical framework of this volume – four main explanatory variables to account for the progress, or lack of progress, in decarbonizing the transport sector:

(1) functional overlap or interrelations between decarbonization goals and the transport sector,
(2) political will,
(3) societal backing, and
(4) the institutional set-up (see Chapter 1).

3.1 Functional overlap

As seen above, quite some policy documents by the European Commission express specific targets for the transport sector and suggest ways forward to achieve them. Because of the broad nature of the transport sector and its many links to other economic sectors such as energy, construction and agriculture, other policy goals can motivate GHG emission reduction in the transport sector and these emission reductions can

contribute to other policy goals. This section discusses the way different policies overlap with respect to emission reductions in the transport
sector and explores potential rebound effects that may occur.

A first evident overlap is with policy goals for energy production.
As electricity will be used more extensively to power light duty road
vehicles, the full GHG emission reduction potential will only be
achieved once electricity is produced from carbon-neutral and renewable sources. Reaching the EU emission reduction targets for transport
will therefore partly depend on the efforts of the power sector to reduce
carbon emissions by 2050. The ETS already sets a cap on emissions, and
various member states have implemented renewable energy policies (see
also Chapter 3). Hence, further greening of the power sector is essential
to reduce GHG emissions of the transport sector.

With respect to biofuels, competition with food crops for land and
water, and concerns about environmental impacts of indirect and direct
land use change resulting from an increased demand for certain feedstock, are important concerns also expressed by the European Commission in the Low-Carbon Roadmap 2050 (European Commission, 2011b).
Uncertainty regarding future availability of sufficient sustainable feedstock makes it likely that the transport sector will have to compete
with other sectors of the economy for carbon-neutral and sustain-
ably produced biofuels (Hill et al., 2009). Therefore, Smokers et al.
(2012) advise the EU to develop realistic policy strategies for biofuels
in transport, integrated into a broader strategy on the most effective use of biomass, to optimize supply and use of non-food biomass.
GHG assessments should take into account land-use change and biodiversity impacts as well as indirect effects in other industries that
may also require feedstock. In addition, a number of non-GHG-related
impacts should be addressed in the policy development, such as impacts
on fuel and food costs, on the transport sector and on other sectors
that either depend on transport or also use biomass (Smokers et al.,
2012).

A second important overlap occurs with transport infrastructure policy objectives. New infrastructure can have ambiguous effects on GHG
emissions: it might encourage the use of less GHG intensive modes
but can also increase GHG emissions if it increases overall capacity or
reduces travel times. This also holds for the new EU Trans-European
Transport Network infrastructure policy, even though it explicitly
aims to aid achieving the EU's long-term transport objectives, namely
meeting future mobility needs while ensuring resource efficiency and
reducing emissions. However, it is not clear how these two objectives

will be achieved simultaneously, as there are no targets defined for the amount of emission reduction that should be attained through infrastructure policy. Hence, overall GHG emissions might increase due to greater traffic volumes on more efficient transport networks, even when the relative emissions per kilometre are lowered due to modal shift and/or cleaner vehicles. Infrastructure policy should focus on climate-friendly infrastructure developments. Attention should thus be given to undesired rebound effects. Improvements in fuel efficiency of vehicles, and thus lower fuel consumption, might make vehicles cheaper to use, so that efficiency improvements might lead to an undesired increase in the amount of transport (Sharpe, 2009).

Energy taxation policy can have an important influence on GHG emission savings. This is supported by an increasing trend in taxing both CO_2 emissions and energy content, as demonstrated in the EU Energy Taxation Directive (Directive 2003/96/EC).

It should also be noted that transport policy measures motivated by reasons other than climate protection can still lead to GHG emission reduction. Stimulating co-modality for passenger transport in urban areas, by promoting slower modes (cycling and walking) and public transport, is encouraged for a wider range of social, economic and environmental reasons, including the social function of public transport and various co-benefits such as congestion reduction, solving parking problems, and reducing noise and air pollution (Skinner et al., 2010). As it is not always clear where priorities are being placed, this further increases the complexity of reaching GHG emission reduction targets in transport.

Policies to reduce GHG emissions from transport have the potential to contribute to the delivery of other policy goals. In the 2050 Roadmap, accelerating the development and early deployment of electrification, and alternative fuels and propulsion methods for the whole transport system is explicitly linked to other EU sustainability objectives such as the reduction of oil dependence (Kollamthodi et al., 2010), the competitiveness of Europe's automotive industry, and health benefits. Reducing transport's GHG emissions could indeed aid in achieving wider air quality objectives required by European legislation (Air Quality Directive 2008/50/EC on ambient air quality and cleaner air for Europe) as a reduction in the amount of fossil fuels used in the transport sector also reduces the amount of conventional pollutants emitted.

The complexity of all these transport-related policy interactions makes it difficult to assess the overall effect of transport policy actions on climate policy objectives. Although many of these interactions are known, it is clear that not all policies are in harmony with each other and higher levels of policy integration between transport and climate policies are required. The potential rebound effects of certain transport policy options underlines the importance of a coordinated approach and the implementation of complementary policy measures in order to ensure that GHG emission reduction targets are reached.[5] Additional research on the nature and size of potential rebound effects is needed in order to allow policymakers to take the interactions between policy objectives into account and try to ensure further harmony between policies.

3.2 Political will

Achieving long-term GHG emission reduction targets for transport by 2050 requires an efficient framework for transport users and operators that ensures a smooth functioning of and effective competition in the internal market, an early deployment of new technologies, and large investments in new transport infrastructure. As listed above, quite some policies exist or are planned, but these are insufficient. In addition, sometimes initiatives are being weakened, losing part of their reduction potential, such as the temporary suspension of the EU ETS requirements for flights operated in 2012–2016 from or to non-European countries under pressure from non-EU countries,[6] or the adjusted Commission proposal for the Fuel Quality Directive after heavy lobbying from the oil industry, Canada and the United States. The policy gap analysis clearly demonstrated that additional policy action is required using a range of technical and non-technical options in a broader and more coherent policy framework.

In order to stimulate the uptake of non-technical options, Skinner et al. (2010) stress that the efficiency of the transport system needs to be further improved through improved spatial planning, speed enforcement, lower motorway speeds and more fuel-efficient driving. Additionally, a level playing field of taxation across all modes is required, to internalize a range of external costs and to remove existing subsidies. For non-technical options, harmonizing fuel duties and VAT across the modes (at the level currently paid by private road transport) could deliver GHG savings of over 10 per cent and implementing a high CO_2 charge in fuel prices could deliver nearly a 20 per cent reduction compared to BAU (Skinner et al., 2010).

All EU member states have some form of road user charging in the form of fuel taxation, which was originally seen as a means of raising revenue rather than as a transport policy instrument. In some cases, fuel taxation has been raised above inflation to reduce CO_2 emissions from transport or differentiated to encourage cleaner fuels (Smokers et al., 2012). However, user charging that internalizes the external costs of the wider impacts of transport is considered to be the first, best and most economically efficient approach for charging vehicles (van Essen et al., 2012). User charging as an instrument to reduce transport's GHG emissions will, however, only work if there is a sufficient reduction in the overall demand for transport. Smokers et al. (2012) therefore argue that road user charging should be introduced in addition to, rather than instead of, fuel taxation. Road user charging at the member state level is only possible if the schemes are consistent with the Road Infrastructure Charging – Heavy Goods Vehicles Directive (Directive 2011/76/EU).[7] However, the current guideline is significantly weaker than the original proposal. Moreover, implementation by member states is on a voluntary basis, severely limiting its potential for emission reduction. Political will is lacking for a more stringent approach.

In turn, policies to achieve energy efficiency improvements are made difficult because manufacturers are required to invest in (initially expensive) technology, while users benefit from the subsequent reduced fuel consumption ('split incentives'; see Skinner et al., 2010; Hill et al., 2012). Therefore, both push (supply side) and pull (demand side) instruments are required, especially regarding transitional technologies characterized by long lead times and requiring high investments in (energy) infrastructure. Also here, additional policy action is required.

With respect to regulations, studies indicate that of all measures to reduce GHG emissions taken in the EU until 2014, CO_2 emission standards for road vehicles have been the most effective (Schade, 2011). There is, however, no EU policy to regulate CO_2 emissions from non-road modes. Some initiatives are being undertaken in an international context, such as emission-related benchmarks for maritime ships by the International Maritime Organization, but progress is slow.

Given the decarbonization objective, Hill et al. (2012) stress the need for early policy action and well-developed communication. Delay of key policy action most likely leads to the need to accelerate action in later years to catch up, with higher risk, increased costs and reduced feasibility of achieving reduction targets. Policymakers should also avoid limiting action only to measures that are cost effective in the short term, since early introduction of measures with higher abatement costs and longer lead times is also required to deliver the savings needed within the timescale available. Additionally, early introduction of new technology is required for modes with long vehicle lifetimes, and thus long times required for fleet turnovers.

3.3   Societal backing

Understanding public attitudes to measures to reduce GHG emissions is a key element in reaching reduction targets. In practice, although the general public is increasingly aware of the role transport plays in GHG emissions, societal backing for policy measures tackling these emissions is often limited.

On the basis of a literature review, Pridmore and Miola (2011) found that the public first of all needs to understand that there is a need for change and be convinced that the measure selected will be efficient. This holds for pricing measures, alternatives to car based transport as well as new technologies and fuels. This can be achieved by clear, accessible information and education. 'Real-life' examples to increase observability in the form of trials and widespread demonstration schemes can play important roles (Rogers, 2003; Winslott-Hiselius et al., 2009). Pridmore and Miola (2011) also suggest that the acceptance of pricing measures will decrease just before implementation, so strong leadership, political will and careful timing are important elements.

Whether, how and where revenue is raised and spent also significantly affects the acceptability of pricing and tax schemes (Pridmore & Miola, 2011). Acceptability increases, in some cases dramatically, if the revenue is spent on either pull measures such as public transport alternatives (Ison, 2000) or to the direct benefit of car users (for example, revenue allocation to decrease transport taxes) (Schuitema & Steg, 2008). The identification of wider benefits is important, for example, health improvements through increased walking and cycling as benefits associated with road pricing (Banister, 2008).

Trust in those implementing the measure and the fairness of the measure have also been identified as key elements by Pridmore and Miola (2011). People need to trust that they are not acting in isolation (with limited benefit) and that changes in behaviour are in line with wider group values and norms. Although not easy to achieve, a number of mechanisms have been identified to facilitate this, including consistent messages, strong and clear political leadership and transparent and accountable revenue spending.

Individual, subjective factors such as green motives and pro-environmental orientation also play a role in increasing the acceptability of measures. For new technologies, the literature on vehicle choice, however, suggests that environmental benefits receive less consideration by the public in comparison with performance and upfront costs (Pridmore & Miola, 2011).

Overall, the involvement of broader societal actors is required in the policymaking and implementation of a sufficiently ambitious transport policy in line with decarbonization objectives. This involvement is currently mostly absent, but crucial, as lack of acceptability for pricing measures, for example, is probably more caused by public perceptions that these measures are ineffective at reducing environmental problems rather than the negative effects of pricing measures on personal car use (Schuitema et al., 2010). Consulting and engaging with members of the public at the measure design stage can help ensure understanding of the measure and its effectiveness (Pridmore & Miola, 2011). Hill et al. (2012) also stress that a wide range of stakeholders needs to be engaged and involved in the solutions.

3.4　Institutional setup

For transport matters, the Council of Ministers acts by qualified majority voting, so that single member states do not have veto power. Nevertheless, decision-making on GHG emission reduction measures in transport is often difficult and lengthy. This was demonstrated by the review of the aforementioned Road Infrastructure Charging – Heavy Goods Vehicles Directive where economic interests of some member states were pitted against the sustainability goals supported by other member states and the Commission.

Quite some lobby groups, often representing a particular mode, such as the International Road Transport Union (IRU) and the Community of European Railway and Infrastructure Companies (CER) can successfully weigh on the political agenda in the field of transport. Also intense lobbying from the power sector and heavy industries resulted in the weakening of the Commission's proposal for the Fuel Quality Directive. Counterweight is provided by some non-governmental organizations such as Transport & Environment, who campaign for smarter and greener transport in Europe.

Action at the EU level complements action at global, member state and regional/local levels. Hill et al. (2012) suggest setting five-year cumulative GHG budgets for the EU transport sector since this could be an effective way to encourage timely action and minimize total cumulative GHG emissions in the long term. According to Skinner et al. (2010), the following policy instruments are needed at the EU level:

- Regulation of energy efficiency of vehicles and GHG intensity of fuels and energy carriers. This involves developing relevant standards for all vehicles for all modes, the progressive tightening of standards, and development in parallel of policy targeting GHG intensity of fuels and energy carriers.

- Standards and criteria to ensure that alternative fuels and energy carriers deliver GHG emission reduction and avoid other adverse sustainability impacts.
- Economic instruments to internalize the external costs of transport for all modes and the harmonization of pricing policies for transport.
- The elimination of existing hidden subsidies and perverse incentives.
- Support for innovation and the development of new technology.
- Review of EU policy towards the development of transport networks.
- Development of evaluation tools to better reflect GHG emissions.

For some specific modes such as international air transport and shipping, action to tackle emissions (such as setting standards) can best be developed at the global level to ensure uniformity of regulation and fair competition. This requires action in and by global organizations such as the International Civil Aviation Organization and the International Maritime Organization. However, sometimes efforts required for global harmonization slow down the regulatory process in EU, as is the case for aviation.

Additionally, Skinner et al. (2010) identify important complementary policy instruments considered to be within the competence of member states or regional and local authorities and where cooperation between the European Commission and relevant authorities is required to achieve coordinated action and share good practices:

- Harmonizing and lowering speed limits.
- Optimal spatial planning and infrastructure policies for GHG reduction in transport that focus on compact cities, bundling of flows and only a limited extension of road and airport infrastructure capacity.
- Setting the framework for the differentiation of vehicle taxes (purchase, registration and circulation) by CO_2 emissions in national law.
- Developing new business models for transport.

Hence, a wide range of policy instruments on appropriate levels of governance is required to ensure that emission reduction goals for transport will be achieved by 2050.

Conclusions

Since 1990, GHG emissions from transport have demonstrated the highest percentage increase of all energy-related sectors. The lack of progress in reducing GHG emissions in transport indicates that decarbonizing the transport sector constitutes a major challenge for European climate policy. The EU therefore has a range of policies in place in 2014 to lower emissions from the transport sector. However, these are not part of an overarching, coherent strategy or goal and are in themselves insufficient to reach the EU GHG emission reduction targets defined in the Low-Carbon Roadmap 2050 and the Transport White Paper. More significant reductions in GHG emissions from transport are needed in order to reach these ambitious but indispensable targets in a timely fashion.

Scenario analysis has demonstrated that a range of both technical and non-technical solutions are required to achieve EU reduction goals through simultaneously reducing GHG intensity of energy used by the transport sector, improving the efficiency of transport vehicles by both technical and operational means, and improving the efficiency of the transport system. Additional analysis showed that co-benefits among policies in different sectors such as energy and agriculture need to be identified and promoted to realize ambitious climate policy goals in transport. In addition, transport policy measures motivated by reasons other than climate protection can lead to GHG emission reduction, while the reverse can also be the case. It is not always clear where priorities are being placed in different policies. Therefore, additional research on the nature and size of potential rebound effects and on potential co-benefits for GHG emission reduction from multiple motivations is essential in order to enable policymakers to take the interactions between policy objectives into account and promote harmony between policies.

Political will is essential in order to achieve the reduction targets but is currently insufficient. Early policy action and well-developed communication are required to limit costs and achieve reduction targets. Political will is also required to ensure societal backing for sustainable transport measures. Overall, broader involvement of societal actors in policymaking and implementation of a sufficiently ambitious transport policy in line with decarbonization objectives is currently lacking, but crucial. Clear, accessible information and education are required for the general public to understand the need for change and efficient measures should be demonstrated. From an institutional viewpoint, decision-making on GHG emission reduction measures in transport is often difficult and lengthy. Cooperation across different levels of governance (global, EU, member state and regional/local levels) for sharing best practices and ensuring coordinated and effective approaches remains a crucial challenge in order to develop coherent policy measures in transport to reduce GHG emissions.

Notes

1. These figures are explored further in the White Paper on Transport on the basis of the Effective Technology scenario (with low fossil fuel prices) of the Roadmap, which shows a 61 per cent reduction for the transport sector (Hill et al., 2012).
2. The Low-Carbon Roadmap warns that this could lead, directly or indirectly, to a decrease of the net GHG benefits and increased pressure on biodiversity, water management and the environment in general, reinforcing the need to advance in second- and third-generation biofuels and to proceed with the ongoing work on indirect land use change and sustainability.
3. The European Commission hopes to finance this new transport infrastructure by, for instance, bond emissions (the 'Europe 2020 Project Bonds' initiative). Another important source of finance needs to come from internalizing external transport costs (European Commission, 2011d).
4. Full reports and an interactive tool showing potential GHG emission reduction from different technologies and policies are available at http://www.eutransportghg2050.eu/.
5. Regulation or target setting can be used as a complementary instrument to enhance the effectiveness of economic instruments such as a cap and trade system or CO_2 tax, by ensuring that sufficient technological options come to the market (Smokers et al., 2010).
6. This suspension was first enacted for 2012 and subsequently extended for the period 2013–2016 to allow time for negotiations on a global market-based measure applying to aviation emissions.
7. A specific provision of this Directive allows member states to introduce a common system of electronic vignettes. In this respect Belgium, Denmark, Luxembourg, the Netherlands and Sweden have a common system of user charges for heavy goods vehicles above 12 tonnes called the 'Eurovignette' system.

References

Banister, D. (2008) 'The Sustainable Mobility Paradigm', *Transport Policy, 15*(2), 73–80.

European Commission (2008) '20 20 by 2020 – Europe's Climate Change Opportunity', COM(2008) 30.

European Commission (2011a) 'A Resource-Efficient Europe – Flagship Initiative under the Europe 2020 Strategy', COM(2011) 21.

European Commission (2011b) 'A Roadmap for Moving to a Competitive Low Carbon Economy in 2050', COM(2011) 112.

European Commission (2011c) 'Energy Roadmap', SEC(2011) 1566.

European Commission (2011d) 'White Paper – Roadmap to a Single European Transport Area. Towards a Competitive and Resource Efficient Transport System', COM(2011) 144.

European Commission (2011e) 'Roadmap to a Resource Efficient Europe', COM(2011) 571.

European Commission (2011f) *Buying Green! A Handbook on Green Public Procurement*, 2nd edn. (Luxembourg: Publications Office of the European Union).

European Commission (2013) *EU Transport in Figures – Statistical Pocketbook 2013* (Luxembourg: Publications Office of the European Union).

European Commission (2014) 'A Policy Framework for Climate and Energy in the Period from 2020 to 2030', COM(2014) 15.

Geurs, K., Nijland, H. & van Ruijven, B. (2011) 'Getting into the Right Lane for Low-Carbon Transport in the EU', in W. Rothengatter, Y. Hayashi & W. Schade (eds.), *Transport Moving to Climate Intelligence: New Chances for Controlling Climate Impacts of Transport after the Economic Crisis* (New York: Springer), pp. 53–73.

Hill, N., Brannigan, C., Smokers, R., Schroten, A., van Essen, H. & Skinner, I. (2012) 'Developing a Better Understanding of the Secondary Impacts and Key Sensitivities for the Decarbonisation of the EU's Transport Sector by 2050. Final Project Report', http://www.eutransportghg2050.eu/cms/assets/Uploads/Reports/EU-Transport-GHG-2050-II-Final-Report-29Jul12.pdf, date accessed 8 October 2014.

Hill, N., Hazeldine, T., Pridmore, A., von Einem, J. & Wynn, D. (2009) 'Alternative Energy Carriers and Powertrains to Reduce GHG from Transport', http://www.eutransportghg2050.eu/cms/assets/EU-Transport-GHG-2050-Paper-2-Alt-energy-carriers-and-powertrains-21-12-09-FINAL.pdf, date accessed 8 October 2014.

Ison, S. (2000) 'Local Authority and Academic Attitudes to Urban Road Pricing: A UK Perspective', *Transport Policy, 7*(4), 269–277.

Kollamthodi, S., Haydock, H. & Falconer, A. (2010) *Energy Security and the Transport Sector. Project: EU Transport GHG: Routes to 2050?* (Brussels: European Commission).

Pridmore, A. & Miola, A. (2011) 'Public Acceptability of Sustainable Transport Measures. A Review of the Literature', *International Transport Forum Discussion Working Paper, 2011*(20).

Rijkee, A. G. & van Essen, H. P. (2010) 'Review of Projections and Scenarios for Transport in 2050', http://www.eutransportghg2050.eu/cms/assets/EU-Transport-GHG-2050-Task-9-Report-V-Draft.pdf, date accessed 8 October 2014.

Rogers, E. M. (2003) *Diffusion of Innovations*, 5th edn. (New York: Free Press).

Sharpe, R. (2009) 'Technical Options for Fossil Fuel Based Road Transport', http://www.eutransportghg2050.eu/cms/assets/Paper-1-preliminary.pdf, date accessed 8 October 2014.

Schade, W. (2011) 'Transport in the Past and Current Climate Policy Regime', in W. Rothengatter, Y. Hayashi & W. Schade (eds.), *Transport Moving to Climate Intelligence. New Chances for Controlling Climate Impacts of Transport after the Economic Crisis* (New York: Springer), pp. 29–40.

Schuitema, G. & Steg, L. (2008) 'The Role of Revenue Use in the Acceptability of Transport Pricing Policies'. *Transportation Research Part F: Traffic Psychology and Behaviour, 11*(3), 221–231.

Schuitema, G., Steg L. & Rothengetter, J. A. (2010) 'The Acceptability, Personal Outcome Expectations, and Expected Effects of Transport Pricing Policies'. *Journal of Environmental Psychology, 30*(4), 587–593.

Skinner, I., van Essen, H., Smokers, R. & Hill, N. (2010) 'Towards the Decarbonisation of EU's Transport Sector by 2050', Final Report, http://www.

eutransportghg2050.eu/cms/assets/EU-Transport-GHG-2050-Final-Report-22-06-10.pdf, date accessed 8 October 2014.

Smokers, R., Skinner, I., Kampman, B., Fraga, F. & Hill, N. (2012) 'Identification of the Major Risks/Uncertainties Associated with the Achievability of Considered Policies and Measures', http://www.eutransportghg2050.eu/cms/assets/Uploads/Reports/EU-Transport-GHG-2050-II-Task-5-FINAL-28May12.pdf, date accessed 8 October 2014.

Smokers, R., van Essen, H., Kampman, B., den Boer, E. & Sharpe, R. (2010) 'Regulation for Vehicles and Energy Carriers', http://www.eutransportghg2050.eu/cms/assets/EU-Transport-GHG-2050-Paper-6-Regulation-for-vehicles-and-energy-carriers-12-02-10-FINAL.pdf, date accessed 8 October 2014.

van Essen, H., Nelissen, D., Smit, M., van Grinsven, A., Aarnink, S., Breemersch, T., Martino, A., Rosa, C., Parolin, R. & Harmsen, J. (2012) *An Inventory of Measures for Internalising External Costs in Transport. Report for Directorate-General for Mobility and Transport* (Brussels: European Commission).

Winslott-Hiselius, L., Brundell-Freij, K., Vagland, A. & Byström, C. (2009) 'The Development of Public Attitudes towards the Stockholm Congestion Trial', *Transportation Research Part A: Policy and Practice, 43*(3), 269–282.

7
Buildings: Good Intentions Unfulfilled

Elin Lerum Boasson and Claire Dupont

Introduction

The European building stock consumes about 40 per cent of final energy in the EU and is responsible for about half of the CO_2 emissions not covered by the EU Emissions Trading System (BPIE, 2011; European Commission, 2013). Reducing buildings' energy consumption reduces demand for fossil fuel-generated energy, thus contributing to a reduction of greenhouse gas (GHG) emissions and to the achievement of decarbonization.

Policies to reduce the energy consumption of buildings sit within initiatives to improve energy efficiency in the EU. Energy efficiency measures have served a variety of broader purposes over the years – in particular to enhance energy security and address climate change. Policies to improve energy efficiency do not present energy efficiency as a goal in itself, but as a means to achieve other ends. In the same vein, improving the energy performance of buildings can be viewed as a policy tool to contribute to reducing GHG emissions in the EU by 80–95 per cent by 2050 (see Chapter 1).

In this chapter, we argue that measures to reduce the energy consumption of buildings have a disappointing history in the EU, and much needs to improve so that the building sector can fulfil its role in achieving decarbonization. Poor implementation of past policy measures; questions about the competence of the EU to legislate on the energy performance of buildings (or questions of 'subsidiarity': the appropriate governance level for policymaking); and the fragmented nature of the buildings' industry have all contributed to the poor record. Furthermore, the 'rebound effect' associated with energy efficiency (namely, offsetting energy efficiency gains with higher consumption) may partly or fully offset the beneficial effects of the new regulations (EEA, 2012, p. 22).

First, we discuss the development of EU policies that target the energy performance of buildings. We then highlight a number of policy gaps in this field in the context of the 2050 decarbonization goal. How effective are the energy performance of buildings Directives? Has the 2012 Energy Efficiency Directive filled some policy gaps? Next, we assess whether (a) functional overlap; (b) political commitment; (c) societal backing; and (d) institutional set-up can help explain why energy performance of buildings policy has progressed to such a limited extent at EU level. We close by highlighting the importance of overcoming past policy trends and failures, and of moving beyond discussions of subsidiarity to foster genuine political will.

1. Historical developments: Policy layering

Policies to improve the energy performance of buildings aim for buildings that have good insulation, with a highly efficient internal distribution of energy, and low-scale renewable energy installations on site. Such buildings can be labelled low- or zero-energy buildings, depending on whether they consume low or no amounts of energy. Through proper design, energy equipment and energy management, buildings can even be transformed from energy consumers to energy producers (often known as 'energy-plus' buildings or active houses) (Heaps et al., 2009).

The EU has agreed an array of energy efficiency measures relevant for the buildings sector (see Table 7.1). Improving the energy performance of buildings can be considered a win-win solution for several issues, including reducing dependence on imports of energy and improving energy security; reducing energy costs to consumers; and combating climate change through reducing demand for fossil fuels. Scenarios for EU decarbonization to 2050 point to the great importance of improved energy efficiency to achieve a reduction in GHG emissions by 80–95 per cent (EREC & Greenpeace, 2010; European Commission, 2011a; Heaps et al., 2009; WWF, 2007, among others; see also Chapter 1), with buildings playing a significant role.

However, the policy instruments agreed throughout the 1990s and 2000s (see Table 7.1) fall short of their potential. While the Commission has often proposed relatively strong measures, support from an organized Brussels-based NGO and buildings industry coalition has only existed since the mid-2000s. Furthermore, attempts by the Parliament

Table 7.1 Selection of EU Directives contributing to improved energy performance of buildings

Year	Directive	Code
1992	Efficiency for Hot-Water Boilers Directive	92/42/EEC
1993	SAVE Directive	93/76/EEC
2002	Energy Performance of Buildings Directive	2002/91/EC
2006	Energy Services Directive	2006/32/EC
2010	Energy Performance of Buildings Directive – recast	2010/31/EU
2012	Energy Efficiency Directive	2012/27/EU

Note: The Directives listed aim either to improve energy performance of parts of the building (such as through improved efficiency of products and services of the construction industry, household appliances and so on) or to outline an holistic building policy.

to strengthen the Commission's proposals have regularly failed in the face of sustained resistance from member states in the Council – a resistance usually backed with arguments in favour of policymaking at lower governance levels and concerns over implementation costs. Such member state concerns are unrelated to the technical developments of the buildings sector, where highly efficient building methods and materials already exist (BPIE, 2011). The member state reluctance comes forth throughout the historical discussion below, and we see no signs that indicate a shift over time.

1.1 Early attempts at EU energy efficiency policy – the 1970s and 1980s

In the mid-1970s, European countries first engaged in policymaking to reduce energy consumption (European Commission 1974, 1979a), in response to the oil crises of that decade. Improving energy efficiency was regarded as an energy security measure to reduce oil consumption. Most advances in energy efficiency, however, were made in a limited number of member states (including the Netherlands, Denmark, Germany and France) (European Commission, 1979b). The EU level seemed to have little influence on policy development.

Policymaking was stepped up in 1984, when the Commission launched an holistic energy-efficiency strategy for the building sector, arguing even then that this sector was responsible for some 40 per cent of the Community's energy demand (European Commission, 1984). However, the member states rejected the Commissions' proposals for a 'thermal auditing' system, and a common reference building code.

In these early days, little progress was made and energy remained a member state competence.

1.2   From energy security to climate change – the 1990s and 2000s

In the 1990s, policy measures to improve the energy performance of buildings became more part of the response to climate change. In the run-up to the Rio Summit in 1992, the Commission launched a new energy efficiency offensive, the SAVE programme, which offered financial support for energy efficiency projects (European Commission, 1990; 1992a). The Commission also proposed a new draft Directive, encouraging member states to enhance thermal insulation and to introduce regular inspection of boilers (European Commission, 1992b). The Council eventually adopted a 'Directive to limit growth in carbon dioxide emissions by improving energy efficiency', the SAVE Directive (93/42/EEC). This Directive did not place any specific obligations on member states.

EU policy remained very weak and, by the mid-1990s, most member states (except Denmark, Sweden and Germany) had scaled back their energy efficiency policies as the oil crises and energy security threats associated with them faded from memory. By the late 1990s, the European Commission was repeatedly arguing that the EU's Kyoto commitment taken on in 1997 required more stringent energy efficiency policy and began work on a new policy proposal for the energy performance of buildings (European Commission, 1998; 2000).

The fragmented nature of the building-construction industry posed a particular challenge for policy development. The Commission was unable to have an extensive industry dialogue, except with some representatives from the insulation industry. Hence the Directive proposal was primarily shaped by the Commission, taking further and specifying core features of the SAVE Directive (ENDS, 2001). The key term in the draft, 'the energy performance of buildings' was a newly coined term that reflected the cognitive shift among building construction experts – away from regulating the energy performance of specific building components and towards regulating holistic energy use.

Throughout the 2000s, member states continued to question EU policy on the energy performance of buildings based on subsidiarity arguments (Boasson & Wettestad, 2013). The Council generally weakened policy instruments, by watering down commitments and/or insisting on longer implementation timelines, as can be seen with the Commission's draft proposal on the energy performance of buildings.

The move to holistic policymaking resulted in the 2002 Energy Performance of Buildings Directive (EPBD: 2002/91/EC). Despite the reluctance of member states (except Denmark) to delegate policymaking authority to the EU (Boasson & Wettestad, 2013), EU energy ministers reached informal agreement on the Directive after only half a year, proposing a few changes and postponing the implementation deadline to 2006 (ENDS, 2002). The EPBD aimed to reduce the energy consumption of buildings by applying energy performance standards for new buildings and for renovations of buildings over 1,000 m^2. Member states could choose to develop certificate schemes that affected the market pricing of buildings, but they were free to design certificate systems that promoted specific technical improvements.

Implementation of the EPBD was disappointing. Although Commission officials continued to argue that 'energy efficiency is a key part of Europe's response to climate change and security of supply issues' (ENDS, 2008), by 2006, only five member states had transposed the EPBD properly (European Parliament, 2008, p. 5). Most member states had adopted energy efficiency targets and support schemes, but with varying levels of ambition. By 2009, only a handful of member states had introduced energy certification of buildings schemes (European Commission, 2009, p. 14).

In 2006, the EU adopted a Directive on Energy End-Use Efficiency and Energy Services (2006/32/EC) to create a framework for energy efficiency policies. This proposal was again severely watered down during the legislative process (Oberthür & Roche Kelly, 2008). The Directive introduced some overarching requirements. Member states should adopt national indicative energy savings targets and develop national energy efficiency action plans. The Directive did not introduce new policy elements to the 2002 EPBD, other than underlining the importance of enhancing the energy performance of public buildings and encouraging energy performance contracting in existing buildings.

In 2006, the Commission launched an energy efficiency action plan where it proposed to introduce a common minimum performance standard for new and renovated buildings, and to develop a strategy for widespread deployment of very low-energy or passive houses (European Commission, 2006, p. 12). Member states were not particularly enthusiastic, again emphasizing that the subsidiarity principle should be respected (European Council, 2006, pp. 4–5). By the mid-2000s, the construction industry and environmental NGOs had begun to organize themselves better at EU level to influence the policymaking machine, but this did not alleviate member states' subsidiarity concerns. Member states were later also concerned about the costs of implementing measures to improve the energy performance of buildings, especially given the financial and economic crises facing the EU since 2008.

The Commission published a proposal for a recast of the 2002 EPBD in 2008 (European Commission, 2008). This was negotiated alongside the broader package of climate and energy policy measures (see Chapter 1). The EPBD recast proposal signalled an increased focus on renewable energy sources, a mandatory EU-wide formula for calculating 'cost optimal energy performance' levels and the removal of the 1000 m^2 threshold in the building code requirements for building renovations. Energy performance certificates were proposed for use in all sales and rental advertising. Overall, although more policy measures were adopted in the 2000s, these measures were low on ambition and poorly implemented.

1.3   EPBD recast and EED – the 2010s

As of 2014, the 2010 recast of the EPBD (2010/317/EU) is the key piece of legislation governing the energy performance of buildings, but the Energy Efficiency Directive (EED: 2012/27/EU) also contributes to the larger picture. Negotiations on both of these policy instruments were fraught with opposition from some member states, again based on questions of competence and subsidiarity.

While the Parliament proposed amendments to strengthen the Commission's original EPBD proposal (by calling for a special energy efficiency fund and arguing that buildings be 'zero-energy' by 2020), the Council was far more reluctant to accept far-reaching measures. The insulation industry was in line with the Parliament's amendments (Boasson & Wettestad, 2013; ENDS, 2009a).

The Council regarded the Parliament's amendments as 'overly ambitious and unrealistic', but the negotiation atmosphere shifted during summer 2009, apparently related to the approaching climate summit in Copenhagen (European Council, 2009, p. 3). A final deal was struck in mid-November (ENDS, 2009b) resulting in recast Directive 2010/31/EU. The core features were in line with the original proposal from the Commission, although some concessions were given to the Parliament. The Parliament's 'zero-energy building' wording was replaced by the ambiguous 'nearly zero-energy buildings'. The agreement said little about financing: member states were merely required to report on their activities in this respect.

The main elements of the recast EPBD are as follows. First, it presents a holistic approach to calculating the *absolute* energy use of the building. A key concept here is 'cost optimal', defined as 'the energy performance level which leads to the lowest cost during the estimated economic lifecycle' (Art. 2.14), which makes costly measures appear more economically viable. Second, the EPBD requires member states to apply minimum standards to new buildings, to buildings that are subject to major renovation, and to technical building systems (Art. 1). Member states must also inspect heating and air-conditioning systems, and independent control systems are required (Articles 14, 15, 17). Third, by 2020, all new buildings are to be nearly zero-energy and new public buildings must fulfil this requirement by 2018 (Art. 9). Fourth, member states are encouraged to introduce financial incentives to catalyse improvements in energy performance. Every three years, member states shall draw up lists of existing and planned financial instruments. Based on the input from member states, the Commission will prepare a report on the use of EU funds and, if appropriate, submit proposals to the European Parliament and the Council (Art. 10). Fifth, the EPBD requires member states to create a system for energy performance certificates, which shall indicate the energy performance of a building calculated in accordance with the methodological requirements of the Directive (Art. 11), and should affect the market price. Member states are to issue certificates for all buildings or building units that are constructed, sold or rented, or have useful floor area of over 500 m^2 (to be lowered to 250 m^2 in 2015) (Art. 12). Independent control systems are mandatory, in order to ensure the credibility of these certificates (Art. 18). The certificates must also include recommendations for improvements. Hence, energy certification is not only a market measure: it also provides people with technical guidance on how to enhance the energy performance of their buildings.

One of the main criticisms of the EPBD recast is that it provides too few incentives to speed up the renovation of the EU's buildings stock. About 40 per cent of the EU-27's building stock was built before 1960. From 1990 onwards, European buildings have been renovated at a rate of about 1 per cent per year on average in the EU (BPIE, 2011; European Commission, 2013, pp. 6–7), but there is very little data about any improvement in energy performance as a result of these renovations. The Buildings Performance Institute Europe suggests that renovation rates of at least 2.5 per cent per year between 2010 and 2050 are required to achieve renovation of nearly 100 per cent of the building stock, provided such renovations lead to buildings consuming close to zero energy

(BPIE, 2011, p. 109). This would reduce the EU's energy consumption of buildings from 40 per cent of final energy today to close to zero in 2050, thereby removing buildings as an energy consumer, as a GHG emitter and as a cause of climate change. The lack of clear incentives for deep and speedier renovations of the EU's existing building stock is one major gap in the EPBD. The EED provides a first policy response to this gap.

The Commission presented its proposal for the EED on 22 June 2011 (European Commission, 2011c). Negotiations centred around three main sticking points. First, the Parliament called for binding national targets to improve energy efficiency, while member states were opposed to this. Second, a renovation target of 3 per cent per year for all public buildings was supported by the Parliament, while member states aimed to reduce the remit of buildings covered by this measure. Third, a 1.5 per cent annual energy savings obligation for energy suppliers was also a point of controversy. Member states wished to provide flexibility for energy suppliers to achieve the target, including by staggering implementation over a number of years, or by including past efforts to count towards the target (ENDS, 2012a; 2012b). The job-creating potential of energy efficiency measures was emphasized by lobbyists and policy-makers in favour of the EED, in the context of a prolonged economic crisis.

The EED (2012/27/EU) was finally adopted after a compromise in trialogue negotiations (informal negotiations among representatives of the three deciding institutions in the EU – the Commission, the Parliament and the Council). Member states are required to set 'indicative national energy efficiency targets', schemes and programmes (Recital 13, Art. 3.1), but no national binding targets. Public building renovation is to take place at the rate of 3 per cent per year, but only central government buildings fall within the scope (Art. 5). Interestingly, governments are required to set down renovation roadmaps for their entire stock of buildings (Art. 4). Finally, energy sales to customers have to achieve 1.5 per cent savings per year to 2020 (Art. 7), but a number of flexible measures to count towards this target are included.

The most important developments for the buildings sector under the EED are the requirements for member states to develop a long-term strategy to mobilize investment in the renovation of all buildings (Art. 4), and the immediate exemplary role to be played by the renovation of central government buildings (Art. 5). These obligations represent the first binding legislative requirements related to the renovation of buildings. This partly addresses the major weakness of the EPBD, which provided few incentives for renovation.

2. Policy gaps

Despite the 'policy layering' described in the historical discussion above, gaps remain if buildings are to fulfil their potential contribution to achieving decarbonization. Reducing the energy consumption of buildings in the EU from 482 mega tonnes of oil equivalent (Mtoe) in 2010 to close to zero by 2050 requires ambitious retrofitting and far-reaching policies, especially considering that energy consumption of buildings has remained at high levels of over 450 Mtoe between 2000 and 2010, even with the EPBD in place (European Commission, 2011b; Eurostat, 2014). Policy measures for buildings should push for high levels of deep renovation (that is, renovations of all aspects of the building to achieve nearly zero energy consumption), and provide incentives or requirements for all new buildings to achieve the highest energy efficiency standards possible.

Furthermore, technical solutions to improve the energy performance of buildings exist, and shortfalls in implementation are more likely due to lack of training among building industry professionals and lack of awareness or financial incentives for consumers (Heaps et al., 2009). Policy measures to fill these gaps would greatly aid the rollout of the existing technologies.

While the EED has made some steps to address the issue of renovating the EU's building stock, both the EPBD and EED lack clear, binding commitments and transfer little authority to the EU. Moreover, the key policy concepts are unclear and ambiguous, like 'nearly zero energy' and 'cost optimal'. Even 'energy efficiency' and 'energy performance' improvements are contested concepts and neither Directive provides clear prescriptions on how to measure achievements and improvements.

The EU has attempted to respond to criticisms about the financial implications of energy efficiency policy measures by aligning a number of funding mechanisms to support projects. Examples of such funding mechanisms include a European Energy Efficiency Fund (EEEF), launched in 2011, providing favourable loans to local and regional authorities for energy efficiency projects. Cohesion policy sets aside more and more funding for energy efficiency projects, with 5.1 EUR billion funding in 2007–2013 for improving energy efficiency. Energy efficiency also forms part of the Horizon 2020 research framework and is a component of structural funds. These funding mechanisms (and others) provide loans and grants for energy efficiency projects where investment would not necessarily flow from the private sector (especially due to the high upfront costs of many energy efficiency measures), and many of the projects funded are small-scale (Rezessy & Bertoldi, 2010). Responding with further and more substantial funding mechanisms may assist in dealing with the policy implementation shortfall, but this strategy does not necessarily alleviate the insufficiently ambitious policy measures agreed.

According to the Commission, the EU is likely to miss its indicative energy savings target of 20 per cent compared to business as usual for 2020 under current policies (European Commission, 2014). The EEA argues that the target can be achieved if further measures are implemented (EEA, 2014, p. 75). Moreover, the Commission highlights that there is a risk of implementation delays of the 2010 EPBD requirements and that this may hamper improvements further. The degrading EU-Russian relations since the 2013 Ukraine crises and the desire to reduce dependence on Russian gas have, however, increased the focus on energy efficiency (see Chapter 10). The European Council adopted a non-binding objective of 27 per cent improvement in energy efficiency for 2030 compared to projected levels of energy consumption, to be reviewed by 2020 (European Council, 2014, p. 5). This was lower than Commission proposals for a 30 per cent efficiency target after the review of the EED in mid-2014. Future revisions of legislation may lead to the adoption of more binding energy requirements for buildings.

3. Analysis: Many measures, weak policy

The Commission has pushed for more binding EU regulation of the energy performance of buildings, with limited success. Policy development follows a pattern where the Commission proposes a certain level of ambition; Parliament attempts to strengthen the ambition of the Commission's original proposal; yet Council weakens the final agreement. Member states introduce new policy elements, often based on unclear and ambiguous concepts, providing more flexibility in the implementation. The resulting policy gaps have often been attributed to member state reticence and to the fragmented nature of the buildings industry in the EU. In the following, we will discuss to what extent the continuing policy gaps may be explained by functional overlap, political will, the involvement of societal interests and the institutional set-up of this policy area.

3.1   Functional overlap

The carbon intensity of energy used in buildings is key to the decarbonization goals of the EU. Reducing buildings' energy consumption and converting remaining energy demands to renewable energy sources will clearly benefit the objective of reducing GHG emissions. We can therefore say that the functional overlap between energy performance of buildings policy and decarbonization objectives is synergistic – reducing energy consumption of buildings helps achieve decarbonization aims.

But energy performance of buildings policies have several other policy co-benefits. Many energy efficiency measures are profitable in the long term, so achieving the policy objectives should in theory be straightforward. However, several decades of energy efficiency policy experience illustrate that few actors are fully economically rational in this area. Further, many actors involved in building construction would profit from enhancing energy performance – refurbishing the European building stock implies increased demand for building material, such as triple-glazed windows, insulation and better ventilation systems. Thus, energy efficiency improvements are expected to provide jobs and boost the economy (European Commission, 2014, pp. 5–6).

However, there are several features that hamper policy development in this issue area. First, there is no consensus on what energy efficiency is and how it should be measured. Is it about reducing the overall energy use or about de-coupling economic growth from energy use? Second, there is an overlap between measures that target GHG emissions in general and measures that target buildings. The EU Emissions Trading Scheme may eventually create higher electricity prices and shifts towards less carbon intensive energy sources (although the scheme has, until 2014 at least, resulted in far lower carbon prices than expected).

Furthermore, the EU's energy efficiency and renewable energy policies are intertwined. It can be argued that the more successful renewables are, the less need there is for energy efficiency, since the main objective is to reduce emissions. Policy emphasis on on-site renewable energy in EU policy for buildings is growing: if it is more profitable to reduce emissions through boosting renewables than with efficiency measures, one should opt for the former. In practice, however, it is hard to know in advance what the optimal levels of energy efficiency improvements and renewable energy production may be. Some authors may argue that until the EU's electricity grid is capable of handling flexible renewables, heightened emphasis on reducing energy consumption may be required (see Chapters 3 and 4).

In sum, energy performance of buildings policy overlaps synergistically with a number of policy areas, including energy security, jobs, competitiveness of energy prices and decarbonization. The emphasis on one policy goal or another may depend on the context within which policy is being developed. During the discussions on the EPBD recast, emphasis was laid on the climate credentials of the policy in the run-up to the international climate summit in Copenhagen in December 2009. In 2014, discussions on the future climate and energy framework have highlighted the energy security dimensions of improved energy efficiency through reduced dependence on energy imports from third countries (especially Russia) (European Commission, 2014). Despite these benefits, policies on the energy performance of buildings have consistently lagged behind on ambition and implementation. The framing of policy goals may be part of the story. It seems clear that understanding the functional overlaps between energy performance of buildings policy and decarbonization objectives does not provide a full picture of why buildings policy has yet to fulfil its potential.

3.2 Political will: Reluctant member states

Throughout the 2000s, political will to take action on climate change was evident in the amount of policies agreed and in the political statements and targets agreed at the EU level (Dupont & Oberthür, 2012). Although targets to 2020 for both renewable energy and reducing GHG emissions were binding on member states, the target to improve energy efficiency was not. This non-binding efficiency target will possibly be unachieved, even with the agreement on the EED in 2012 (European Commission, 2014). So how can the EU demonstrate relatively high levels of political will to act on climate change, yet consistently produce weak or poorly implemented policies on the energy performance of buildings?

In this case, member states themselves seem to be to blame. The Parliament has consistently called for stronger policy measures in energy performance of buildings, and it can be seen as the most ambitious of the EU's institutions in this field. The Commission has put forward new proposals to fulfil the potential of energy savings. Political attention has tended to produce more EU regulations in this area, but many are ambiguous and non-binding. EU member states have repeatedly drawn on the subsidiarity principle as a justification for high flexibility in implementing policy measures. Member states have thus argued that the EU is not the appropriate level for policymaking on the energy performance of buildings.

In the Council, it is possible to distinguish three groups of member states in terms of their positions on the energy policy for buildings. The first group includes Denmark, Sweden and Germany, which developed ambitious building codes and extensive public funding quite early (Boasson & Wettestad, 2013). These countries have tended to be at the forefront of EU policy development and often strengthened their national policies in advance of EU decisions. Germany distinguished itself by introducing requirements to promote on-site renewable energy before EU policy measures were adopted. Later, Austria (with its high focus on passive houses) and the United Kingdom (with its specific focus on CO_2 abatement) joined the frontrunners when it came to national efforts. By 2009, all the leading countries had decided that new buildings constructed after 2020 (2030 for Austria) were to be low-energy houses (Thomsen & Aggerholm, 2009; Schild et al., 2010, p. 22). Despite these significant national achievements, not all of these governments became eager supporters of a stronger EU policy in this issue area.

The second group of countries included most of the other 'old EU member states' with little tradition of energy efficiency policies (Belgium, Finland, France, Greece, Ireland, Italy, Luxembourg, the Netherlands, Portugal and Spain). They implement EU policies in this issue area rather reluctantly (see Papadopoulou et al., 2009; Molina & Álvarez, 2009). The third group consists of Central and East European countries that only started to develop energy efficiency policies after entering the EU. Much of the building stock in these countries was in urgent need of repair and improvement (Peterstorf et al., 2005). Boasson and Wettestad (2013) highlight that many of these newer member states liberalized their housing market, but lacked detailed regulation and organization of building ownership. This dispersed responsibility for energy-related investments, and made it hard to develop appropriate regulation.

The two latter groups, but also some member states among the frontrunners, have repeatedly emphasized that EU energy policy for buildings should be based on the subsidiarity principle (see European Council, 2006, pp. 4–5). In addition, the energy performance of buildings has a rather low to modest political significance in most EU countries. Few member states or national parliaments paid much attention to the political deliberations on the first EPBD (Boasson & Wettestad, 2013). This issue area has always been lower on the political agenda than other climate policy measures, such as the Emissions Trading Scheme and renewables. While political backing for action on climate change may be high in most EU member states, this political will is not found in buildings policy, with questions of competence trumping potential policy developments.

3.3   Societal backing: Business and civil society

EU citizens have consistently indicated their support for policies that combat climate change, including on energy efficiency (Eurobarometer, 2008; 2011; 2014). In 2014, Eurobarometer reported that 92 per cent of people surveyed thought that it was important for their government to provide support for improving energy efficiency by 2030 (Eurobarometer, 2014, p. 57). If the continued support for energy efficiency policies among EU citizens is well considered, we could perhaps expect higher levels of ambition among policymakers. Such societal backing is often also communicated through lobby groups or other stakeholders at the EU level.

Policies that target the energy performance of buildings are of relevance to many societal groups, but we will primarily focus on the construction industry and environmental organizations. The building industry can be divided into three groups: constructors, building product producers and building managers. The construction industry includes architects, construction contractors, plumbers, carpenters, roofers and so on. It is a very heterogeneous group, as each of the different professions tends to specialize, with most employed in specialized firms that represent one profession only. The construction industry is the largest industry employer in Europe, but most firms are small and local. The European Construction Industry Federation (FIEC) estimates that 95 per cent of the construction firms in the EU area have fewer than 20 employees (FIEC, 2010). Not many companies are publicly traded. Firms generally focus on surviving in the business and operate with fairly short planning horizons (Boasson, 2015). These actors are typically represented in Brussels by small business associations that consist of a range of national associations. Large transnational construction contractors (like German Hochtief, French Vinci Construction and Swedish Skanska) have no direct representation in Brussels. This group has not been much involved in influencing EU policy development (see Boasson & Wettestad, 2013).

The second group, the building product producers, includes industrial actors like cement, steel, glass and insulation. Each sub-industry here is dominated by fairly concentrated industries, comprising of 5–15 large enterprises with a European or global outreach (see, for example, Nordqvist et al., 2002).[1] Most firms are publicly traded. However, whereas the insulation and glass industries create products that are applied only or primarily in buildings, many of the others make products used for many purposes. That means that few product producers were involved in the EPBD processes to any extent, with the exception of Eurima, the insulation trade association, and EuroACE, an organization that promotes a more ambitious EU energy policy for buildings (ENDS, 2000; EuroACE, 2011).

The third group consists of building managers. They include commercial landlords, firms and organizations that ensure public, cooperative and social housing, and all the individuals and small/medium firms that own their own building. Professional owners did not get a Brussels presence until the European Property Federation was established in 1997 (EPF, 2011). Public, cooperative and social housing organizations are represented through a common organization, Housing Europe. Individual owners are not represented in Brussels. Overall, building managers do not have a strong presence on the Brussels scene (see Boasson & Wettestad, 2013).

Thus, these industry groups have not engaged strongly in developing the EU energy policy for buildings. The fragmented and national, or even local, nature of this industry and the lack of presence on the Brussels energy efficiency scene mean that they have only modest impact on policy development.

Environmental NGOs generally support stronger energy efficiency policies. However, many NGOs came late to the energy efficiency issue. It was not until the mid-2000s that the major NGOs dedicated staff and resources to the issue. Lack of resources was likely a barrier for environmental NGOs to push policy forward. From the mid-2000s onwards, several NGOs accessed policymakers at the EU level with relative ease. However, their influence was limited rather to the Commission and the Parliament, as such NGOs often find it difficult to influence policy negotiations in the Council or in trialogue discussions (Dupont, forthcoming).

With both industry groups and environmental NGOs facing difficulties in ensuring member states agree on sufficiently ambitious policy, they also have decided to join forces at times. As of 2014, the coalition for energy savings includes 27 industry and environmental groups that present a common message to EU policymakers in favour of ambitious policy on energy efficiency in general.[2] Although joining their voices may help strengthen the importance of the message, member states remain unconvinced of the benefits of ambitious EU-level policy on the energy performance of buildings.

In sum, it can be said that EU citizens, and societal groups, including many segments of industry and environmental NGOs, are positive towards ambitious policy on the energy performance of buildings. This positive societal backing has yet to be translated into ambitious policy measures on the energy performance of buildings, with member states being the most important blocking or reluctant actors.

3.4 Institutional set-up: National traditions creating path dependency

We can define two different strands of path dependency influencing this policy area: the traditional approaches of member states and the position of the Commission. These historical policy traditions show the preferences of many member states to keep energy policy within their competence, and efforts of the Commission to gain more influence in the energy sphere.

First, many countries have long national histories of regulation of energy performance in buildings, and have their traditional methods. The United Kingdom and Germany, for example, traditionally regulate buildings at a regional or local level. Member states have been reluctant to transfer authority to the EU because they have regarded this as a technical issue for which they need to develop solutions that fit within their usual working methods and their specific climate conditions. National/local building regulations regulate a lot more than energy, with issues such as safety and fire-prevention as other major concerns.

Second, we see a remarkable stability in the positions of the Commission from the mid-1980s and onwards. Commission officials have held onto the objective of developing EU wide building regulations, and have also promoted the certificate market measure. Policy proposals from the Commission tend to respond to the weak implementation of earlier policy measures. Later proposals try to ensure previous policy is finally implemented or to strengthen the remit or scope of the measures already in place. The limited competence of the EU in this issue area hinders the Commission from proposing strong and ambitious policy measures.

However, not all Commission officials are enthusiastic about strong building measures. There has been a long-standing discussion between market supporters within DG Climate Action and DG Energy arguing that any measures that might interfere with the CO_2 price mechanisms of the ETS should be avoided (see Boasson & Wettestad, 2013). It seems that the challenges in this sector will not readily be resolved with changes in energy prices. Achieving behaviour change in favour of energy efficiency can be hampered by consumption systems, structural factors and the rebound effect (EEA, 2013). Lack of expertise on how to improve the energy performance among building constructors and building owners requires development of targeted policies.

EU policy on the energy performance of buildings falls within the ordinary legislative procedure, where the European Parliament and Council must agree on a proposal from the Commission before a piece of legislation is adopted. Therefore, decisions in the Council can be taken under qualified majority voting (QMV), preventing a single member state veto, which in theory could help prevent serious weakening of policy proposals. However, in the cases of the EPBD and EED, the majority of member states were in any case unenthusiastic about such policies. They demonstrated little interest in working together on the issue of energy efficiency and QMV did not prevent the Council watering down the proposals. This could be because buildings are not commodities that are tradable across borders, pose little impact on competitiveness across the EU, and there is thus limited incentive for member states to create a 'single market for buildings' (again raising subsidiarity concerns).

The Parliament is somewhat caught between the Commission, that constantly proposes policies that take small steps forward, and the Council, that does not see the real benefit for EU policy in this area. Historically, the Parliament has been considered an 'environmental champion' (Carter & Burns, 2009; Rasmussen, 2012). In the case of policy on the energy performance of buildings, the Parliament has fulfilled this role by proposing amendments that are more ambitious and far-reaching. The Parliament has rarely managed to push through its most ambitious and stringent amendments, however. In this case, it is clear that although the policy falls under the ordinary legislative procedure, it is the Council that retains most clout in the tri-institutional relationship (as it plays the easier 'blocking' role).

In general, the institutional set-up of this policy field at the EU level highlights the differences among the three institutions. The Commission may continue to propose policy and the Parliament will probably try to strengthen these policy measures, but the Council has the final word and may continue to water down the provisions. These roles have hardly shifted since energy efficiency policy first entered the EU policy field in the mid-1980s.

Conclusions

While the EU began developing its energy efficiency policy for buildings before climate change became a central policy issue, competence at the EU level to expand on this policy has remained limited. A poor record of implementation in member states is testament to the low level of commitment to such policies. The portfolio of EU policies keeps growing, but the actual ability to steer the actions of member states remains constrained. These challenges can be linked to the questions of subsidiarity and the multitude of actors in the sector. If decarbonization in the EU is to be achieved, the buildings sector must move quickly to implement ambitious performance standards for new and existing buildings, and effective policies must drive the sector in this direction.

From the analysis above, two points become clear. First, political will to promote improved EU policies for energy performance of buildings is not evident in member states although societal backing from EU citizens, industry and environmental NGOs has grown. Second, past policy developments, institutional traditions and poor implementation seem to result in new policies that try to catch up on past failings rather than put in place adequate new policy measures. Changing these two elements at the political level ought to be part of future policy developments.

Generating political will to agree on ambitious policy on the energy performance of buildings may require more than usual lobbying from industry and environmental NGOs. Until the many co-benefits of improved energy performance of buildings are fully understood by policymakers, the issue may not receive the attention it deserves. It is clear that energy performance of buildings policy can serve several policy objectives, and clever policy framing to ensure ambition is required. This may require continuous training of policymakers, and ongoing research into the many positive effects of energy performance of buildings policies on society. Education for consumers (the electorate) may also tip political pressure towards more stringent policy measures.

Moving beyond a pattern of fixing the failures of past policy measures may be more problematic. This requires efforts to increase the engagement of industry and NGOs, and calls for impartial research on the effect of adopted measures. This discussion is challenging given different views on appropriate instruments, differences in definitions and understandings. Definitions of 'high energy-performance buildings' flourish, including concepts like 'passive houses', 'low-energy buildings', 'fossil-fuel independent houses' or 'zero-energy buildings' (European Commission, 2009; Schild et al., 2010, p. 22). Further, it is hard to imagine that the building construction industry will become a more coherent force on the European scene anytime soon: the industry is simply very decentralized and fragmented in nature. We still have a lack of good and comparative mapping of all the various national activities. A discussion on whether the EU is the right level for development of regulatory measures for this sector may prove valuable, given the continuous subsidiarity arguments. If it is, then what form should that policy take? If it is not, how can the EU level ensure national measures are comparable for the purposes of accounting the reduction of GHG emissions? EU funding mechanisms providing financial assistance to local and regional authorities may prove one step towards resolving concerns over costs of policies and the appropriate level for implementation, but these do not respond to the problem of the EU level agreeing insufficient policy ambition.

It may be difficult to imagine that changes in political will in member states will be possible and that the adoption of new policy measures for ensuring the energy performance of buildings improves. Even a political crisis like the one between Russia and Ukraine from 2013 onwards proved insufficient to push EU leaders to adopt binding targets for energy efficiency to 2030 (European Council, 2014, p. 5). Further incentivizing of energy efficiency measures, such as through linking with other benefits received through EU funding or policy mechanisms, may prove more fruitful for advancing energy efficiency rather than continued reliance on EU policy agreements. Other windows of opportunity will need to be seized in future, if buildings are to play their role in reducing consumption of fossil fuels.

Notes

1. See EURIMA, www.eurima.org/headquarters/, date accessed 9 February 2012, and Glass for Europe, www.glassforeurope.com/en/about/our-members.php, date accessed 17 June 2011.
2. See energycoalition.eu for details, date accessed 26 September 2014.

References

Boasson, E. L. (2015) *National Climate Policy: A Multi-Field Approach* (London: Routledge).

Boasson, E. L. & Wettestad, J. (2013) *EU Climate Policy: Industry, Policy Interaction and External Environment* (Farnham: Ashgate).

BPIE (2011) *Europe's Buildings under the Microscope: A Country-by-Country Review of the Energy Performance of Buildings* (Brussels: Buildings Performance Institute Europe).

Carter, N. & Burns, C. (2009). 'Is the European Parliament an Environmental Champion?', *Full research report, ESRC end of award report, RES-000-22-2304* (Swindon: ESRC).

Dupont, C. (Forthcoming). *Climate Policy Integration into EU Energy Policy* (London: Routledge).

Dupont, C. & Oberthür, S. (2012) 'Insufficient Climate Policy Integration in EU Energy Policy: The Importance of the Long-term Perspective' *Journal of Contemporary European Research*, 8(2), 228–247.

EEA (2012) *Consumption and the Environment – 2012 Update* (Copenhagen: European Environment Agency).

EEA (2013) 'Achieving Energy Efficiency through Behaviour Change: What Does It Take?' *Technical Report*, No 5/2013 (Copenhagen: European Environment Agency).

EEA (2014) 'Trends and Projections in Europe 2014: Tracking Progress towards Europe's Climate and Energy Targets for 2020', *EEA Report*, No. 6/2014 (Copenhagen: European Environment Agency).

ENDS (2000) *Insulation Firms Slam EU Energy Efficiency Plan*, 10 August (London: Haymarket Media Group Ltd).

ENDS (2001) *EU Aims for More Energy-Efficient Buildings*, 26 April (London: Haymarket Media Group Ltd).

ENDS (2002) *EU Buildings Efficiency Law Agreed*, 3 October (London: Haymarket Media Group Ltd).

ENDS (2008) *Analysis: Prices and Targets Pressure Shifts Focus to Lower Demand*, 1 February (London: Haymarket Media Group Ltd).

ENDS (2009a) *EU Building Energy Law Revision 'Must Go Further'*, 6 January (London: Haymarket Media Group Ltd).

ENDS (2009b) *All New Buildings to be 'Near-Zero-Energy' by 2020*, 18 November (London: Haymarket Media Group Ltd).

ENDS (2012a) *Commission Upbeat about an EED Deal by June*, 8 March (London: Haymarket Media Group Ltd).

ENDS (2012b) *EED Trialogue Talks Set to Be Very Challenging*, 29 March (London: Haymarket Media Group Ltd).

EPF (2011) *European Property Federation*, www.epf-fepi.com/index2.html, date accessed 17 June 2011.

EREC & Greenpeace (2010) *Energy [R]evolution: Towards a Fully Renewable Energy Supply in the EU 27* (Brussels: Greenpeace International and European Renewable Energy Council).

EuroACE (2011) *EuroACE in a Nutshell*, http://www.euroace.org/AboutUs/AboutEuroACE.aspx, date accessed 17 June 2011.

Eurobarometer (2008) *Europeans' Attitudes towards Climate Change* (Brussels).

Eurobarometer (2011) *Climate Change. Special Eurobarometer 372* (Brussels).

Eurobarometer (2014) *Climate Change. Special Eurobarometer 409* (Brussels).

European Commission (1974) 'Community Programme for Rational Use of Energy (RUE)', *Information Memo P-70/74*.

European Commission (1979a) 'Third Report on the Community's Programme for Energy Saving', COM(79) 313.

European Commission (1979b) 'New Lines of Action by the European Community in the Field of Energy Saving', COM(79) 312.

European Commission (1984) 'Towards a European Policy for the Rational Use of Energy in the Building Sector', COM(84) 614.

European Commission (1990) 'Proposal for a Council Decision Concerning the Promotion of Energy Efficiency in the Community', COM(90) 365.

European Commission (1992a) 'A Community Strategy to Limit Carbon Dioxide Emissions and to Improve Energy Efficiency', COM(92) 246.

European Commission (1992b) 'Proposal for a Council Directive to Limit Carbon Dioxide Emissions by Improving Energy Efficiency (SAVE programme)', COM(92) 182.

European Commission (1998) 'Energy Efficiency in the European Community – Towards a Strategy for the Rational Use of Energy', COM(1998) 246.

European Commission (2000) 'Green Paper: Towards a European Strategy for the Security of Energy Supply', COM(2000) 769.

European Commission (2006) 'Action Plan for Energy Efficiency: Realising the Potential', COM(2006) 545.

European Commission (2008) 'Proposal for a Directive of the European Parliament and of the Council on the Energy Performance of Buildings (recast)', COM(2008) 780.

European Commission (2009) 'Low Energy Buildings in Europe: Current State of Play, Definitions and Best Practice', September 2009, http://ec.europa.eu/energy/efficiency/doc/buildings/info_note.pdf, date accessed 7 February 2014.

European Commission (2011a) 'Energy Roadmap 2050', COM(2011) 885/2.

European Commission (2011b) 'Impact Assessment Accompanying the Document: Energy Roadmap 2050', SEC(2011) 1565 Part Two.

European Commission (2011c) 'Proposal for a Directive of the European Parliament and of the Council on Energy Efficiency and Repealing Directives 2004/8/EC and 2006/32/EC', COM(2011) 370.

European Commission (2013) 'Financial Support for Energy Efficiency in Buildings', SWD(2013) 143.

European Commission (2014) 'Energy Efficiency and Its Contribution to Energy Security and the 2030 Framework for Climate and Energy Policy', COM(2014) 520.

European Council (2006) 'Presidency Conclusions', Document 7775/06, 23 and 24 March.

European Council (2009) 'Note from Secretary General to Delegations. Subject: Energy Efficiency Package', Interinstitutional File: 2008/0222 (COD), 2008/0221 (COD), 2008/0223 (COD), 29 May.

European Council (2014) 'Conclusions', Document EUCO 169/14, 23 and 24 October.

European Parliament (2008) 'Compromise Amendments 1–25 on the Proposal for a Directive of the European Parliament and the Council Amending Directive 2003/87/EC', Draft report, 2008/0013 (COD), 5 October.

Eurostat (2014) 'Statistics Online Database', http://epp.eurostat.ec.europa.eu/portal/page/portal/eurostat/home/, date accessed 5 May 2014.

FIEC (2010) *Key Figures. Activity 2009* (Brussels: European Construction Industry Federation).

Heaps, C., Erickson, P., Kartha, S. & Kemp-Benedict, E. (2009) *Europe's Share of the Climate Challenge: Domestic Actions and International Obligations to Protect the Planet* (Stockholm: Stockholm Environment Institute).

Molina, J. L. & Álvarez, S. (2009) 'Spain: Impact, Compliance and Control of Legislation', *ASIEPI Information Paper*, P172, www.buildup.eu/publications/7050, date accessed 13 February 2012.

Nordqvist, J., Boyd, C. & Klee, H. (2002) 'Three Big Cs: Climate, Cement and China', *Greener Management International*, 39(Autumn), 69–82.

Oberthür, S. & Roche Kelly, C. (2008) 'EU Leadership in International Climate Policy: Achievements and Challenges', *International Spectator, 43*(3), 35–50.

Papadopoulou, K., Papaglastra, M., Laskari, M. & Santamouris, M. (2009) 'Evaluation of the Impact of National EPBD Implementation in MS', *ASIEPI Information Paper* P180, www.buildup.eu/publications/7368, date accessed 13 February 2012.

Peterstorf, C., Boermans, T., Joosen, S., Jakubowska, B., Scharte, M., Stobbe, O. & Harnisch J. (2005) 'Cost-Effective Climate Protection in the New EU Member States. Beyond the EU Energy Performance of Buildings Directive', Report by ECOFYS for EurEMA (Brussels: EurEMA).

Rasmussen, M. K. (2012) 'Is the European Parliament Still a Policy Champion for Environmental Issues?', *Interest Groups & Advocacy, 1*(2), 239–259.

Rezessy, S. & Bertoldi, P. (2010) *Financing Energy Efficiency: Forging the Link between Financing and Project Implementation* (Brussels: Joint Research Centre of the European Commission).

Schild, P. G., Klinski, M. & Grini, C. (2010) 'Sammenligning og Analyse av Krav til Energieffektivitet i Bygninger i Norden og Europa', Project report 55, Comparison and Analysis of Energy Performance Requirements in Buildings in the Nordic Countries and Europe (Trondheim: SINTEF Byggforsk).

Thomsen, K. E. & Aggerholm, S. (2009) 'Denmark: Impact, Compliance, and Control of Legislation', *ASIEPI Information Paper* P175, www.buildup.eu/publications/7047, date accessed 14 February 2012.

WWF (2007) *Climate Solutions. WWF's Vision for 2050: Short Topic Papers* (Gland, Switzerland: WWF International).

8
The Geopolitics of the EU's Decarbonization Strategy: A Bird's Eye Perspective

Tom Casier

Introduction

In the twenty-first century, there has been a strong tendency to look at international energy relations in terms of geopolitical power. Control over energy resources or pipelines is seen as a significant geopolitical asset. In this context, the EU plays a peculiar role as a major energy consumer but marginal producer, rendering it dependent on imports. With an ambitious decarbonization objective set for 2050, however, the EU's position and that of its energy suppliers is bound to change. But how? At first sight, if we copy today's dominant images of control over resources, the answer may seem obvious. Countries with many hours of sun, abundant wind, hydropower or biomass could be expected to grow stronger, and traditional suppliers of oil and gas relatively weaker. The reality is far more complex and outcomes are dependent on a wide variety of factors, from commercial choices to technological developments. Even more, the question of geopolitics is not purely a matter of facts and data. It is equally a matter of perception, of framing developments on regional and global energy markets in a wider context.

This chapter explores and reflects on how decarbonization policies may affect the regional and global geopolitics of energy and the EU's position within it. Complex questions need to be answered. How will energy consumption change, not just in the EU, but worldwide? How will the energy mix evolve? How will production evolve and what are the capacities? Who has control over technology, R&D and patents may highly determine or counter patterns of dependence. Energy price setting will be of key importance. How will changing demand and supply affect interdependence patterns? Will old asymmetries be replaced by new asymmetries? What will be the larger political context, determining the understanding of geopolitical strengths and weaknesses? Finally, what will be the regulatory context and who is able to influence it?

This chapter does not aim to answer all these questions or to forecast energy politics in 2050. It seeks to outline a few trends that may drastically alter the stakes of the energy 'game'. I argue that the emphasis is likely to change away from a simple geopolitical reading of control of energy supplies as a source of power towards leadership in technology and regulation. In the next section, I analyse the current focus on energy geopolitics and dependence. The subsequent section lists reasons why it is likely that this geopolitical reading of energy relations loses its predominance in the context of decarbonization policies. I then turn to future energy scenarios with regard to three aspects of (geo)politics: the geopolitics of supply, the politics of technology and the politics of regulation. Throughout the analysis the emphasis is on renewable energy and energy efficiency measures (leading to lower consumption). Nuclear energy is left out because of the uncertainties about its future and the limited role the EU attributes to this form of energy in its own scenarios (see also Chapters 1 and 3).

1. Understanding geopolitics and energy

1.1   The emergence of a geopolitical energy narrative

The focus on resource nationalism and the geopolitical interpretation of behaviour on energy markets is somehow understandable in light of the 'supply crunch', resulting from three different factors (Goldthau & Witte, 2010, pp. 9–10). First, reserves of traditional energy resources that can be exploited at a low cost have been dwindling (though the exploitation of shale gas may reverse this trend). Second, demand for energy is rising rapidly, mainly as a result of emerging economies, in particular China and India. Demand is expected to rise by 40 per cent between 2009 and 2035 (IEA, 2011). Third, investments cannot keep up with rising energy demand. Huge investments are needed, not least in the field of exploration and the development of new technologies. However, 'a singular focus on hard security in global energy alone entails the risk of generating misleading analyses and policy prescriptions' (Goldthau et al., 2010, p. 342).

To see the complexity of energy relations, we need to move away from a narrow focus on energy as a zero-sum game and resource nationalism. Seeing energy relations as an arena of geopolitical competition for control over supplies and pipelines is not the result of hard material facts.

It is also determined by how we understand these facts in a broader context and how we give meaning to the behaviour of various actors. While energy never lost its geopolitical significance, energy relations in Europe in the 1990s were predominantly seen as part of the economic realm. At a time of low energy prices and relative optimism, energy was in the first place a commodity being traded with the purpose of making commercial profit. In 2014, a different political narrative exists parallel to the economic and commercial discourse (Aalto & Westphal, 2008; CIEP, 2004). Energy is to varying degrees seen as a strategic asset, a fundament of power or a potential threat for a country's security. Arguably this has equally become a dominant approach in academic literature (see, for example, Smith, 2010; Paillard, 2010; Baran, 2007). Control over hydrocarbons or over their transmission is seen as an important strategic asset, enhancing the relative capabilities of one state, while creating dangerous dependence for the other. With hydrocarbons becoming scarcer and new economies emerging, competition increases (CIEP, 2004, p. 118; Dreyer & Stang, 2014, pp. 13–14).

This chapter takes a different approach. Political competition is not seen as a given, following from material facts (though material changes are acknowledged). Rather, it looks at geopolitical considerations and frames of reference as they exist in the minds of decision-makers and as they are socially produced and understood (see also Hadfield, 2008). In other words, it is argued that future energy geopolitics will not simply be determined by objective facts, but will be a matter of how changes in energy relations are framed. Energy relations between the EU and Russia illustrate how big the divergence between the 'hard facts' and images of dependence may be. In particular since the gas spats between Russia and Ukraine in 2006 and 2009, the EU's perceived excessive dependence on Russian energy became a major political concern and fostered strategies for energy security. In reality, however, the dependence of EU member states on Russia in relative figures has substantially decreased over the last decades, from a 55 per cent share of Russian gas imports in the EU-27 in 1990 to 31.5 per cent in 2008 (Casier, 2011). To explain this discrepancy, we need to look at the perception and social images, and at the broader context in which these perceptions arise. The same holds for the geopolitics of decarbonization. How a change towards renewables and higher energy efficiency will affect geopolitical relations will not only be determined by changes in import and export or by availability of energy resources, but equally by the understanding of the broader political and normative context, which cannot be predicted.

1.2 Geopolitics and energy dependence

Therefore, I approach geopolitics as a two-step model. The first step is to look at geopolitical relations as specific forms of asymmetrical interdependence. This interdependence can be based on energy supply and demand, but also on technology or infrastructure. The second step is to look at the broader context and to see how images of energy dependence get a certain meaning within this context. In other words, how is energy dependence understood as generating power or as posing a threat?

The first step is based on Keohane and Nye's theory of complex interdependence (1989). In their view interdependence produces 'reciprocal (although not necessarily symmetrical) costly effects of transactions' (Keohane & Nye, 1989, p. 9). If the costs are higher for one party (for example, the importer) than for the other (for example, the exporter) the interdependence is asymmetrical. This asymmetry creates a potential source of influence (Keohane & Nye, 1989, pp. 10–11). Moving beyond Keohane and Nye, I do not approach these costly effects in purely material and rationalist terms (assuming a situation in which actors can calculate their costs and benefits without any bias), but understand them as socially mediated, within a certain ideational context. In other words, what matters is how both parties perceive the costs.

Keohane and Nye further add a crucial distinction between sensitivity and vulnerability. Sensitivity refers to the short-term costly effects, which are generated by the behaviour of one actor. Vulnerability refers to the long-term effects generated. If country A is importing high amounts of energy from country B, it will be sensitive. In other words, if country B stops all supplies to country A, the latter will suffer considerable costs as a result. However, it is only vulnerable if it has no alternative and if the costs are therefore doomed to stay high in the longer term. This distinction is important, as only vulnerability creates a potential strategic advantage for state B.

For carbon fuels, the nature of the energy source determines the degree of vulnerability. Oil is traded on international, flexible markets and predominantly shipped by tankers. This implies that alternatives exist. If an oil exporting country cuts off its exports to an oil importing country, the latter has the chance to buy its oil from a different country on the international market. Prices may go up, but the vulnerability of the importing country is limited. This is very different for gas markets. Natural gas is predominantly transported through pipelines, in particular towards Europe, where LNG represents only 19 per cent of gas demand (IGU, 2013, p. 10) and decreased in 2012 and 2013.[1] Pipeline gas is subject to long-term bilateral contracts, which provide for individual fixed price-setting (linked to oil or coal indexes) and contain destination clauses prohibiting the gas from being re-traded (Goldthau & Witte, 2010, p. 5). If a country is excessively dependent on the import of natural gas, it finds itself in a vulnerable position. If it has no pipeline connections with other countries and no alternatives (for example, prospects for domestic exploitation of shale gas), it is in a vulnerable position. The monopolistic gas exporter holds considerable geopolitical power, because it has the 'potential to affect outcomes' (Keohane & Nye, 1989, p. 11). This potential can exist only if there is no demand vulnerability; that is if the energy exporter is not one-sidedly dependent on demand from one consumer.

For energy dependence to generate power, a second step is required. Supply vulnerability will only generate power if energy relations are seen as the dominant context, overshadowing other relations of power in different contexts. Here, images play a very important role. The importance that is attributed to energy in different power settings is dependent on the context and how it is perceived, not on 'objective' dependence. In 2014, there is a global context of rising energy demand and the absence of an institutionalized environment, but also of images of certain actors and their intentions.

2. How decarbonization will affect the geopolitics of energy

In a carbon world, the dominant images of energy geopolitics are about control: control over supplies and over transit. One's geopolitical power – the capacity to affect outcomes in other states – is seen to be determined by the availability of energy resources and the control over transmission. This control can easily be seen as a zero-sum game. The gains for a state achieving control over fossil fuels or over pipelines is an equal loss for its counterpart, in particular in a context where energy resources become scarcer.

The issue of control is different in the case of most renewables under a decarbonization scenario. Wind and sun, for example, are common goods. They are available in all countries, albeit to varying degrees. Some states may have considerable benefits, for example because they are exposed to substantially more hours of sunshine, but they are unlikely to achieve a monopolistic or even dominant position on the market. Other countries in the same geographic area will usually share the advantages. If the EU wants to import electricity, generated by solar energy, from Northern Africa, it can seek collaboration with multiple countries. Decarbonization will thus lead to a less oligopolistic context of energy suppliers.

A second major difference between today's carbon world and tomorrow's decarbonized world, is that renewable energy is in most cases not transported similarly to fossil fuels. Coal, oil, gas are shipped or transported through pipelines and burned in the country of demand. In most cases (the uncertain market of bio-fuels being an exception), low-carbon energy (hydro, solar, wind, tidal, geothermal power, biomass) is transformed into electricity on the spot, in the country of production. The electricity is then transmitted to the countries of consumption. This implies that the availability of electricity infrastructure, storage capacity and transborder interconnections become of crucial importance.

This has a few implications. The importance of regional energy relations (as opposed to global) is likely to increase. As costs increase with distance, electricity will be imported predominantly from countries in proximity. Furthermore, infrastructure will be of crucial importance. Are connections between different electricity networks available? Are they technically compatible? Who controls the networks and who may decide to interrupt deliveries? How are interconnections and transmission regulated? Both technical and legal aspects (who sets the standards?) will be strongly determining. Related to those aspects, we may see regional energy clusters gaining in importance, both in terms of interconnectedness and in terms of common technical and legal standards.

Third, technological changes are taking place at spectacular speed and constantly change positions of relative energy power and dependence (as the impact of shale gas in the United States has demonstrated). This holds even more for the renewables sector, where new technologies have a huge impact on efficiency and cost-effectiveness. Strengths will be determined to a high degree by access to the most performing and cost-effective technology.

Finally, decarbonization may lead to more decentralization of energy production. Energy may be produced in smaller quantities at local level for local consumption. As a result self-sufficiency may increase. The 'Energiewende' in Germany is a good case in point (Bosman, 2012). Similar effects may result from energy efficiency measures. The EU Roadmap 2050 expects that 'centralised large-scale systems such as e.g. nuclear and gas power plants and decentralised systems will increasingly have to work together' (European Commission, 2011a, p. 8).

In sum, the geopolitical relations in a decarbonized world are likely to be less oligopolistic and zero-sum driven. Regional relations will gain importance. Strengths and weaknesses will be less determined by access to energy resources and more by access to technology and infrastructure. Finally, control over regulatory standards is expected to be of decisive importance. The former elements imply that some important geopolitical effects will be indirect rather than direct. Infrastructure, technology and common regulation will lead to wider variation of energy prices in different regions of the world. This will have an indirect impact on investment, on competitiveness and growth.

Of course, we are far from this ideal–typical scenario of a decarbonized world. With rapidly rising energy demand, the geopolitics of fossil fuels will continue to dominate energy relations for many years to come. In this transition context it is unlikely that low-carbon countries will be able to translate their limited dependence on fossil fuels into a 'potential to affect outcomes'. However, it is likely that the position of current conventional energy producers will be affected by their capacity to diversify their energy export. In other words, will they be able to reinforce their positions by converting into major players in the non-fossil fuel market as well?

3. Geopolitical implications of the 2050 decarbonization scenarios

3.1 Towards a post-carbon era? Projecting future energy relations

Will the decarbonization of energy be a geopolitical 'game-changer' (Mirtchev, 2013)? Forecasting energy relations is an extremely difficult exercise. Determining patterns of change, empowerment and opportunities depends on how different factors intersect. First, it is necessary to look at patterns of consumption and production. Both will depend on economic growth, but also on new energy efficiency technologies. They will affect import and export flows and the capacity for self-sufficiency. Second, the degree to which patterns of demand and supply in different regions enter into competition with each other is important and affects price setting. Third, political and ideological choices made in energy and economic policies play a role. This concerns climate targets, decisions on investments in renewables, public funding of decarbonization and so on. One of the most uncertain factors is the choice for nuclear energy, with some European countries having opted out because of safety concerns. Fourth, the energy future depends on technological developments and their (un)even distribution throughout the world. Fifth, the scarcity of conventional resources that may be exploited at low costs may affect energy dependence in the longer term. Finally, investments will determine future energy positions and relations.

The energy mix is changing rapidly. Shale gas and unconventional oil are on the rise and have substantially affected the position of the United States, moving from a net energy importer to energy exporter. But the fastest growth is in the renewables sector (IEA, 2013), signalling a greening of economies throughout the world, although this development is uneven and occurs at different paces. This greening has long moved beyond Europe and it is now countries like China who invest most in renewable energy (IEA, 2013, p. 5). Nuclear energy also continues to rise, mainly in China, India, Korea and Russia and despite the withdrawal of some countries (IEA, 2013). Because of the difficulty of this exercise, most scenarios of future energy relations are predominantly based on an extrapolation of existing trends, integrating new policy targets.

To understand the geopolitics of the EU's decarbonization policy, the analysis relies on existing scenarios of decarbonization forecasting trends up to 2050 (European Commission 2011a), as presented in Chapter 1. All high decarbonization scenarios point to a reduction of energy consumption in the EU by 2050, mainly because of higher efficiency. They suggest that EU energy consumption will be a combination of renewables and to a lesser extent conventional energy sources. Renewables, predominantly wind and biomass, are likely to account for more than half of the consumption. While the consumption is expected to decline in absolute figures, the relative share of natural gas is more or less stable (around 20–25 per cent). The shares of oil and solid fuels drop considerably (halved to 15 per cent in the case of oil). As to nuclear energy, there is uncertainty about possible opt outs (European Commission, 2011a). Figures also suggest a strong electrification: electricity is expected to double its share in energy demand to 36–39 per cent in 2050 (European Commission, 2011a, p. 6). The Commission expects the share of renewables in electricity generation to continue to increase, attaining at least 43 per cent in 2030 and 50 per cent in 2050 (European Commission, 2013, p. 43). It is within these parameters that the subsequent analysis will be done.

3.2   Changing patterns of energy dependence

Under EU decarbonization scenarios, import dependency of the EU-28 is expected to drop from 58 per cent to 35–40 per cent in 2050 (European Commission, 2011a, p. 5). Relative import dependence on natural gas, however, will moderately increase, even with declining consumption.

This is the result of the limited availability of gas within the EU. With fossil fuels taking a smaller share of the energy mix, the EU would be less sensitive to volatile prices of fossil fuels on the international market. Wind energy is mainly produced within the EU, increasingly off-shore. Even under the less optimistic scenarios, biomass is 'expected to be mainly indigenous beyond 2020' (European Commission, 2013, p. 50). When it comes to electricity, there is a continuous decrease of electricity net imports, with negative figures as of 2015, growing consistently until 2050 (European Commission, 2013, p. 86).

But we also need to take into account that dependence itself may be framed differently because the 'material' energy context will be reshaped fundamentally. First, EU energy relations will be more regional than today because the share of oil is expected to decrease considerably. Oil is traded on the international markets and potentially shipped over long distances. In gas trading, LNG may be a game changer, but seemed past its peak in 2012 after years of consecutive growth. In the EU, LNG represents only one-fifth of the gas market. Natural gas is therefore likely to stay (although in lower absolute amounts), but is mainly dependent on regional pipelines. Together with the electrification of energy supply, this implies that infrastructure will be of paramount importance and will become much more central to energy policies. Regional grids and interconnections will be key to energy security. The overall outcome is the increasing importance of regional connections and partnerships, though global consumption patterns will continue to affect these regional links and price setting.

Second, we are likely to see new regional energy players as a result of the rising share of renewables. North-African countries, for example, have a considerable potential when it comes to solar energy. However, new players in the renewables market may not play a dominant role. The main renewables will be wind and biomass (European Commission, 2013). Both are expected to be produced predominantly domestically, while the share of imported solar energy will be limited.

Third, as gas continues to be a relevant energy source, the conventional gas suppliers (Russia, Norway and to a lesser extent Algeria) may continue to play an important role. On the one hand, with gas production in the EU decreasing, gas imports may rise modestly. On the other hand, overall gas consumption is expected to decrease because of higher energy efficiency. But there are two major factors of uncertainty. First, the share of gas in the 2050 energy mix will depend on the development of commercially viable technology for Carbon Capture and Storage (CCS). Second, the gas market is changing rapidly and deeply, not least because of developments in the exploitation of shale gas. This may result in EU member states becoming new gas producers (such as Poland). Also neighbouring transit countries (such as Ukraine) may become suppliers of gas. Moreover, the EU will no doubt seek to further diversify its gas imports and reduce the vulnerability of individual member states by creating interconnections between national gas networks. Overall, dependence on gas imports may remain roughly unaltered under decarbonization scenarios if CCS is successful. It continues to be the source of energy which potentially creates the strongest form of dependency, though gas markets are slowly becoming more flexible (Goldthau & Witte, 2010), in particular as the producers of natural gas are likely to face increasing competition of LNG suppliers. Finally, as global trade in gas is expected to double between 2009 and 2035, external trends in consumption will affect the gas market considerably (IEA cited in Oettinger & Novak, 2013).

Fourth, some of the conventional energy suppliers may reinforce their role if they are able to diversify their energy exports into the renewables sector. Russia has a huge potential in the field of biomass. Norway may contribute through wind energy. Algeria may play a new role through solar energy.

In sum, with a lot of renewable energy produced domestically, EU import dependence is bound to decrease substantially. The biggest 'losers' will be the oil states one-sidedly dependent on oil for their energy export (see also Rothkopf, 2009; Burrows & Treverton, 2007). If decarbonization in emerging economies elsewhere in the world lags behind, they may be able to compensate this loss by rechanneling their exports. In the case of effective CCS, the biggest 'winners' will be the conventional gas suppliers to the EU, who could obtain a solid position in the field of renewables as well. However, with electricity playing a crucial role, the nature of dependency is bound to change. Unlike gas, electricity will not be subject to long-term contracts and rigid price setting, but will be a more flexible market with fluctuating prices and a high degree of competition. Again, this makes good interconnections vital: avoiding energy islands is a key objective.

3.3   Impact on EU energy partners

How does this translate to individual energy partners of EU countries? The image is rather diversified. Some core energy suppliers to the EU, like Russia, Norway and to a lesser extent Algeria, are likely to continue to play a key role. Several OPEC countries remaining too dependent on oil exports, may see their roles change drastically. Finally, new energy suppliers are emerging, in particular in North Africa (see Table 8.1).

Table 8.1 Summary of the impact of decarbonization on the role of EU energy partners

Country	Current role	Potential role	Challenges
Russia	– Key exporter of conventional fuels	– Continuing role as gas exporter – Huge potential in renewables	– Unlocking renewables potential – Infrastructure and electricity interconnections – Different regulatory frameworks
Norway	– Key exporter of gas, oil and green electricity	– Continuing role as gas exporter – Huge potential as green electricity exporter	– Few (close political relations with EU, proximity and good interconnection) – Domestic petroleum industry
Persian Gulf states	– Key oil exporters	– Declining role unless diversification	– Diversification away from oil exports – Limited renewables potential (distance)
Algeria, Libya, Morocco	– Algeria: important exporter of gas and oil – Libya: significant oil exporter and transit country – Morocco: marginal role	– All: considerable renewables potential – Algeria: continuing role as gas exporter	– Unlocking renewables potential – Stable investment climate – Infrastructure and electricity interconnections
Turkey	– Important transit country	– Further increase of gas transit – Potential gas producer – Potential in renewables	– Unlocking renewables potential – Further integration into EU regulatory framework

Table 8.1   (Continued)

Country	Current role	Potential role	Challenges
Azerbaijan	– Important oil and gas producer, but limited export to EU	– Considerable potential for gas export	– Infrastructure interconnections
Ukraine	– Important transit country	– Continuing role as gas transit country – Potentially own gas production	– Infrastructure interconnections – Integration into EU regulatory framework – Relations with Russia

Source: Author.

Russia will continue to be an important energy partner of the EU (European Commission, 2011b; 2011c; see also Chapter 10). However, despite growing export markets in Asia, the contribution of oil and gas to the Russian GDP is expected to decrease from one quarter to 15 per cent of its GDP in 2035 (IEA, 2011, p. 69). In the gas sector 'interdependence is likely to remain a key feature of the EU-Russia energy relationship in the coming decades' (Oettinger & Novak, 2013, p. 10). However, there remain many uncertainties in EU-Russia gas relations. In particular, prices may heavily impact on EU gas demand in a context of increased energy diversification (Oettinger & Novak, 2013, p. 13). In general, Russia is planning to increase its gas production, diversify its exports (Eastward), invest in LNG and in the construction and renovation of pipelines.

Russia is striving to obtain a 4.5 per cent renewables in its energy mix by 2020, but does not seem to be on track to meet this objective. If we leave out hydropower, renewables represent 1 per cent of its energy production only (Lee, 2011). While short-term incentives for investing in the renewables sector are small in a country with abundant fossil fuels, its potential for renewable energy is huge, in particular in the field of biomass and wind. One of the main challenges is the country's connection to the European grid.

The combination of declining oil exports, a huge potential in renewables and the need for better electricity interconnections may in the longer term create a common ground for closer cooperation between the EU and Russia. The EU-Russia Roadmap sets a strategic target to achieve by 2050 'a Pan-European Energy Space, with a functioning integrated network infrastructure, with open, transparent, efficient and competitive markets' (Oettinger & Novak, 2013, p. 5). The EU-Russia Energy Roadmap mentions a 'significant potential to increase electricity trade between both sides' (Oettinger & Novak, 2013, p. 7). Russia is investing in increasing the capacity of its 'Unified Power System' (a synchronous transmission grid dating back to the Soviet Union connecting Russia to several CIS states and to the Baltic states) and in better interconnections with its neighbouring countries. There is potential for EU-Russia cooperation on renewable technology, but this is inhibited by different regulatory systems, difficult access to markets and legal uncertainty related to investments in Russia (see Chapter 10).

Relations with neighbouring countries will also affect opportunities for cooperation. Ukraine, for example, is an important transit country for Russian gas, but also strongly dependent on Russia. Its transit role has come under pressure as a result of the Nord Stream pipeline, directly connecting Russia and Germany overseas. It continues, however, to have a strong potential as interface between the EU and Russia. Its potential for shale gas may reduce its energy dependence in the future.

In 2008, Norway accounted for 15 per cent of EU oil imports and 30 per cent of natural gas imports (European Commission, 2011b). This renders the country a key and reliable energy partner for the EU. As a European net oil exporter it has invested some of its oil income in renewables, like wind energy and hydropower (Müller-Kraenner, 2007, p. 113). Norway has the long term potential to act as Europe's 'green battery' through its 'pumped-storage hydropower', but, as of 2014, progress has been slow and incremental (Gullberg, 2013, p. 615). Norway scores particularly well when it comes to infrastructure (EY, 2013, p. 14), and its proximity to the EU gives it a considerable advantage. Its oil industry is doomed to lose in the longer term, but this is expected after a peak only. Because of its role as gas exporter and its potential as green electricity exporter, Norway is predestined to remain a key partner for the EU (see also Chapter 11).

The Caspian Sea basin forms both a resource-rich area and a crucial transmission hub (see Chapter 9). It is also in the vicinity of Iran, another major energy producer, and of Turkey. It is an area where different players believe they have important interests: Russia, the United States, Iran, Turkey and European states. The Caspian region came to be seen 'not only as a critical component of Western energy security, but also as a linchpin in the evolving balance of power in Eurasia, Asia and the Middle East' (Heslin, 1997). The US initiative to construct the Baku-Tbilisi-Ceyhan oil pipeline has set a geopolitical logic of post-Cold War energy relations into motion, explicitly aiming to access the richness of the Caspian Sea basin, while bypassing Russia and Iran (Heslin, 1997).

Azerbaijan is a traditional and major oil exporter, but accounts for only 3 per cent of EU oil imports (European Commission, 2011b, p. 3). As a consequence, the impact of EU decarbonization policies in the longer term will depend in the first place on the knock-on effect on other countries and the effect on international oil prices. Potential losses could be partly compensated by gas exports. Azerbaijan is home to one of the biggest gas developments worldwide (the Shah Deniz field) and has more gas reserves offshore in the Caspian Sea (EIA, 2014). As a result it has become a net gas exporter and may play a new role as energy provider in the future. Georgia is an important transit country, with both the Baku-Tbilisi-Ceyhan (oil) and the South Caucasus pipelines (gas) crossing its territory. In this role it functions as an important interface between Azerbaijan and the West. On the other side of the Caspian Sea, Turkmenistan is a very important gas producer. However, it is dependent on Russia's pipelines and its energy sector is interwoven with the Russian one. Gazprom has a long-term contract with Turkmenistan, reselling its gas at much higher prices (Müller-Kraenner, 2007, pp. 48–49).

Turkey has considerable potential as a transit country linking the EU on one hand and the Caspian Sea basin, Iran and Iraq on the other. In 2014, it is already a strategically important transit country, both because of the naval route through the Bosporus Strait (through which 3 million barrels of oil per day are shipped; EIA, 2014) and because of pipelines. Overall, Turkey's gas market is expected to grow, overtaking oil, but also to undergo substantial change. Depending on political developments, Iran's share in gas imports may increase. While its gas production is currently marginal, Turkey has some gas reserves itself (around 6 bcm – billion cubic metres) and is likely to increase its production. Moreover, new gas supplies discovered off the coast of Cyprus and Israel open the perspective of new transit routes, though both may face considerable political challenges.

So far Turkey has focused more on gas than on renewables, where it has a long way to go to develop its full potential. The country has significant hydropower capacity and varied potential for other renewables, such as wind, solar and geothermal energy. It is ranked 24th in the EY renewable energy country attractiveness index (EY 2013, p. 14).[2] Its position in a regional, more decarbonized energy market is expected to increase in importance. It is also significantly enhanced by its status as candidate EU member state, which implies that the energy *acquis* gets transferred and funding is channelled towards the energy sector (Dreyer & Stang, 2014, p. 43).

Northern Africa has two important energy exporters already: Algeria and Libya. Their future depends on the long term availability of gas and oil. All Maghreb countries have an enormous potential in solar energy. They have an average sunshine of more than 300 days per year and the desert provides the necessary space for solar plants (Müller-Kraenner, 2007, p. 137). Algeria has been an important exporter of natural gas and LNG to the EU. It accounts for 15 per cent of EU imports of natural gas (European Commission, 2011b). It is also the second largest African oil producer and an OPEC member. Algeria has a good chance of reinforcing its position as a regional energy player if it successfully develops its renewable potential. The Algerian government plans to produce 40 per cent of domestic electricity consumption from renewables by 2030 (EIA, 2014). Morocco is also attempting to develop its role on the renewables market, with high potential for both wind and solar energy. It scored reasonably well in the 2013 EY renewable energy country attractiveness index (EY, 2013, p. 14), in 32nd position. If the country develops successfully, it may become a newcomer on the energy market. Libya, an OPEC country, is an important oil exporter and transit country. It is assumed to have large reserves. Ninety-eight per cent of its export revenue comes from energy, predominantly oil (IMF quoted in EIA, 2014). Declining demand for oil in the longer term may be compensated by gas exports and a new emphasis on renewables. In both sectors, the country has great potential, but much investment is required before it can be exploited. The close link between political turmoil and energy resources places risks on the development of the gas and renewables sectors.

In general, some countries in the Middle East and Northern Africa (MENA) have the potential to become significant exporters of renewably generated electricity. A good grid connection and investment climate form key conditions. The difficult road to realizing this renewable energy potential was illustrated by the fate of the DESERTEC consortium. This group of German industrial companies had to abandon its prestigious project to build a 400 billion EUR solar plant in North Africa in 2013. The withdrawal of important partners from the consortium (Siemens and Bosch in 2012, E.ON in 2014) was ascribed to technological changes and regional instability obstructing investments.

In the longer term the biggest victims of the EU's decarbonization policy may be the traditional oil exporters from the Persian Gulf. If global oil demand were to fall, in the wake of EU decarbonization, they would see revenues dwindle and their exceptional political position would change. This is likely to be reinforced by the United States becoming a net exporter of energy and by increasing levels of energy self-sufficiency in other countries. Emerging economies may somehow make up for that loss, because of increasing demand, but this is likely to be temporary. Countries of the Persian Gulf have a strong potential in solar energy, but are disadvantaged by their geographic position at relative distance from the main consumers. Some countries can be said to anticipate changes and seek to reduce one-sided dependence on oil export. Qatar, for example, holding the world's third largest gas reserves, has become the global leader in LNG.

3.4   The new determinants of energy politics: Infrastructure, technology and regulation

As argued above, the real stakes of energy politics in a decarbonized scenario may be less about control over resources than over good infrastructure, connections, access to technology, and rules and standards.

With electricity doubling its share by 2050 (European Commission, 2011a, p. 6), infrastructure will be a highly determining factor for energy relations. Good interconnections between the EU and countries in its geographic proximity will determine whether the potential will be realized. Ultimately they will be very determining for a country's position in relations of interdependence and for its geopolitical role. Infrastructure requires investments, government steered policy and bilateral agreements. The right decisions at the right time will determine whether the potential of non-carbon energy resources can be unlocked.

For the EU, setting up the right flexible infrastructure is a key objective: 'An overall increase of interconnection capacity by 40 per cent up to 2020 will be needed, with further integration after this point' (European Commission, 2011a, p. 15). Strong emphasis is put on connections with the Sahara and with Russia (European Commission, 2011c).

Access to technology and the use of cost-effective technology will also become much more determining than in the past. Research and development, as well as intellectual property rights will be crucial. The impact on interdependence should not be underestimated. In certain cases, the relation of dependence between energy consumer and producer may shift or transfer into a win-win partnership. If European companies hold patents of new, efficient technologies in solar energy, for example, they may enter into a partnership with North-African companies and governments to set up a high-tech solar plant to produce electricity for export to Europe.

Both infrastructure and technology are interrelated with the third crucial determinant of new energy geopolitics: regulation. The future politics of decarbonized energy relations will to a large extent be determined by systems of governance of international energy markets and by who is setting the standards and thus enjoying regulatory hegemony. Given that we may expect a bigger proliferation of energy sources and energy producers, one of the core challenges becomes the flexible and cost-effective transmission of energy, in particular of electricity. This requires multilateral institutional arrangements on the regulation of markets, investments and public actors and environmental and technical standards. In 2014, there is a complex patchwork of institutional arrangements, both at regional and global level (Dreyer & Stang, 2014, p. 29).

The importance of multilateral energy market regulation is bound to increase for different reasons. First, the need for longer-term investments and for more flexible trade in energy will necessitate the opening of markets to foreign capital and thus more international agreement on the terms and conditions. This may put pressure on energy producers to move away from their traditional preference for (state) control over production and transmission and lead to more liberalized and flexible markets (Kaveshnikov, 2010).

Second, growth in international institutional arrangements for governing energy markets may be driven by general developments in trade liberalization. We have entered a new stage of trade liberalization, moving beyond the abolition of tariffs towards deep and comprehensive free trade arrangements. These imply the harmonization of regulations and standards in fields far beyond classic free trade matters (technical, safety, environmental standards; intellectual property rights; investment regimes; and so on). Setting the standards gives a key competitive advantage in the longer term. Other countries will face adaptation costs. In the field of economic regulation of energy markets, the EU has actively tried to establish this regulatory hegemony through the Energy Community and through Eastern Partnership policies.

Setting the international standard is equally important in the field of climate change. The three different energy narratives – on the market, security and the environment (CIEP, 2004; Aalto & Westphal, 2008; Bressand 2012) – have become more and more intertwined (Nutall & Manz, 2008). In other words, the international arrangements and procedures to tackle climate change will increasingly determine the political choices and the competitive positions of other countries.

To conclude, the complex system of energy governance in the field of liberalization, market regulation, investment, climate change and other environmental rules may ultimately determine much more who will 'win' or 'lose' in energy politics than having access to specific resources. The EU, as an institutionalist and regulatory actor par excellence, and as the biggest trading bloc in the world, is in a pole position at this point. But whether it will be able to keep its leadership role will depend on many factors, not least the capacity to engage in global and regional partnerships.

Conclusions

The current one-sided focus on the geopolitics of energy relations follows from rapidly increasing global consumption, the increasing depletion of low-cost reserves, high investment needs and uncertainties about changing patterns in production and consumption. However, energy relations are determined by a complex interaction of various players and interdependencies with both contrasting and overlapping dynamics: commercial, regulatory, environmental, political, technological and so on. Moreover, a geopolitical understanding of energy relations is ultimately not just a matter of control over resources and energy transmission – and the asymmetrical interdependencies they generate – but as much of images and projections exogenous to the raw energy data. Understanding the energy politics in a decarbonizing world is therefore a difficult balancing exercise – one that can highlight possible trends rather than make precise forecasts. Energy relations will depend not just on consumption and production of energy, but on a multitude of factors, including investments, price-setting, market positions and so on.

Drawing on the EU's 2050 decarbonization scenarios, this chapter argued that the EU's future energy environment may be much more regional and diversified. Energy resources and players will likely be more proliferated, and part of the renewable energy production will be domestic. The regional aspect follows from the importance of infrastructure, in particular the grid that allows the transmission of electricity generated by renewable sources to the EU. The potential available in neighbouring countries can only be exploited if there is sufficient, compatible and flexible infrastructure connecting these countries with the EU. Contrary to oil, gas is expected to play a continuing role, even if consumption in absolute figures may decrease. In particular, in the case of cost-effective CCS, the conventional gas suppliers of the EU will likely remain important partners. Potentially, they may even reinforce their position if they manage to combine this with developing their potential in renewable energy. Russia has an enormous potential to do this, but has a long way to go in terms of investments and infrastructure.

The proliferation of actors and energy sources may make a narrow geopolitical reading in terms of control over resources less likely and the EU less vulnerable. Two other factors may shape the energy politics of the future. First, the control over technology will be crucial. Access to new cost-effective technology for renewables will determine a country's comparative advantage. Second, the political wrestling will likely be more over regulatory standards and institutional governance of energy. Getting your norms, rules and institutional practices accepted in such diverse fields as climate change, trade, investment or taxation is key. Who is setting the rules on energy and climate change may eventually determine a country's position more strongly and – with a wider array of energy sources and producers at hand – maybe even more than the pure availability of resources. For both technology and regulation, it is about the comparative advantage that follows from being in a leading position of setting the standard or controlling patents. As the biggest trading bloc with a tradition of rule expansion beyond its borders, the EU is not in a bad starting position for an energy future in which technology and regulation may play a key role, but its position is not guaranteed. It will require a coherent energy policy, internally and externally, as well as investments in research and development of new technologies. Finally, flexible regional and global partnerships will be important, to guarantee both solid interconnections and leadership in setting international rules. But ultimately, the political reading of this future energy constellation will depend on other factors and political images unrelated to energy.

Notes

1. LNG deliveries to the EU went down 31 per cent in 2012, mainly because of high Asian LNG prices (Market Observatory for Energy, 2013, p. 1). Worldwide, LNG represents roughly one third of gas trade (BP, 2013).
2. EY assesses a country's renewable energy attractiveness on the basis of five parameters, reflecting business opportunities in the sector, not the actual share of renewables in a country. See http://www.ey.com/UK/en/Industries/Cleantech/Renewable-Energy-Country-Attractiveness-Index—Methodology.

References

Aalto P. & Westphal, K. (2008) 'Introduction', in P. Aalto (ed.), *The EU-Russian Energy Dialogue: Securing Europe's Future Energy Supply* (Aldershot: Ashgate), pp. 1–21.

Baran, Z. (2007) 'EU Energy Security: Time to End Russian Leverage', *The Washington Quarterly, 30*(4), 131–144.

Bosman, R. (2012) 'Germany's "Energiewende". Redefining the Rules of the Energy Game', *Briefing Papers* (The Hague: Clingendael International Energy Programme).

BP (2013) *BP Statistical Review of World Energy. June 2013* (London: British Petroleum).

Bressand, A. (2012) 'The Changed Geopolitics of Energy and Climate and the Challenge for Europe', *CIEP Paper, 2012*(4).

Burrows, M. & Treverton, G. F. (2007) 'A Strategic View of Energy Futures', *Survival: Global Politics and Strategy, 49*(3), 79–90.

Casier, T. (2011) 'The Rise of Energy to the Top of the EU-Russia Agenda: From Interdependence to Dependence?', *Geopolitics, 16*(3), 536–552.

CIEP (2004), *Study on Energy Supply Security and Geopolitics: Final Report* (The Hague: Clingendael International Energy Programme).

Dreyer, I. & Stang, G. (2014) 'Energy Moves and Power Shifts: EU Foreign Policy and Global Energy Security', *Issue Report*, No 18, February.

EIA (2014) 'Countries' – Diverse country profiles from the US Energy Information Administration, http://www.eia.gov/countries, date accessed 6 May 2014.

European Commission (2011a) 'Energy Roadmap 2050', COM(2011) 885/2.

European Commission (2011b) 'Key Facts and Figures on the External Dimension of the EU Energy Policy', SEC(2011) 1022.

European Commission (2011c) 'On Security of Energy Supply and International Cooperation. "The EU Energy Policy: Engaging with Partners beyond our Borders" ', COM(2011) 539.

European Commission (2013) *EU Energy, Transport and GHG Emissions. Trends to 2050, Reference Scenario 2013* (Luxembourg: Publications Office of the European Union).

EY (2013) *RECAI: Renewable Energy Country Attractiveness Indices, August 2013*(38).

Goldthau, A. & Witte, J. M. (2010) 'The Role of Rules and Institutions in Global Energy: An Introduction', in A. Goldthau & J. M. Witte (eds.), *Global Energy Governance* (Berlin: Global Policy Institute; Washington, D.C. Brooking Institution Press), pp. 1–21.

Goldthau, A., Hoxtell, W. & Witte, J. M. (2010) 'Global Energy Governance: The Way Forward', in A. Goldthau & J. M. Witte (eds.), *Global Energy Governance* (Berlin: Global Policy Institute; Washington, D.C.: Brooking Institution Press), pp. 341–356.

Gullberg, A. T. (2013) 'The Political Feasibility of Norway as the "Green Battery" of Europe', *Energy Policy, 57*, 615–623.

Hadfield, A. (2008) 'Energy and Foreign Policy: EU-Russia Energy Dynamics', in S. Smith, A. Hadfield & T. Dunne (eds.), *Foreign Policy: Theories, Actors, Cases* (Oxford: Oxford University Press), pp. 321–338.

Heslin, S. (1997) 'The New Pipelines Politics', *The New York Times*, 10 November 1997.

IEA (2011) *World Energy Outlook 2011* (Paris: International Energy Agency/OECD).

IEA (2013) *World Energy Outlook 2013* (Paris: International Energy Agency/OECD).

IGU (2013) *World LNG Report* (Oslo: International Gas Union), http://www.igu.org/sites/default/files/node-page-field_file/IGU%20-%20World%20LNG%20Report%20-%202013%20Edition.pdf, date accessed 24 April 2014.

Kaveshnikov, N. (2010) 'The Issue of Energy Security in Relations between Russia and the European Union', *European Security, 19*(4), 585–605.

Keohane R. & Nye, J. (1989) [1977] *Power and Interdependence* (New York: Harper Collins).

Lee, A. (2011) 'Country Profile, Russia: A Thaw in Official Attitudes Could Rouse Renewable Energy's "Sleeping Giant"', *Renewable Energy World*, http://www.renewableenergyworld.com/rea/news/article/2011/03/country-profile-russia, date accessed 25 April 2014.

Market Observatory for Energy (2013) *Quarterly Report on European Gas Markets, 6*(1), http://ec.europa.eu/energy/observatory/gas/doc/20130611_q1_quarterly_report_on_european_gas_markets.pdf, date accessed 24 April 2014.

Mirtchev, A. (2013) 'The Greening of Geopolitics', *European Energy Review*, 6 May 2013.

Müller-Kraenner, S. (2007) *Energy Security. Re-Measuring the World* (London: Earthscan).

Nutall, W. J. & Manz, D. L. (2008) 'A New Energy Security Paradigm for the Twenty-First Century', *Technological Forecasting & Social Change, 75*, 1247–1259.

Oettinger, G. & Novak, A. (2013) 'Roadmap EU-Russia Energy Cooperation until 2050', March 2013, http://ec.europa.eu/energy/international/russia/doc/2013_03_eu_russia_roadmap_2050_signed.pdf, date accessed 24 April 2014.

Paillard, C. A. (2010) 'Russia and Europe's Mutual Energy Dependence', *Journal of International Affairs, 63*(2), 65–84.

Rothkopf, D. J. (2009) 'Is a Green World a Safer world? A Guide to the Green Geopolitical Crises Yet to Come', *Foreign Policy*, September–October, 134–137.

Smith, K. (2010) 'Russia-Europe Energy Relations: Implications for US Policy' (Centre for Strategic and International Studies), http://csis.org/publication/russia-europe-energy-relations, date accessed 21 June 2011.

9
Decarbonization and EU Relations with the Caspian Sea Region

Claire Dupont

Introduction

In this chapter, I explore the potential consequences and opportunities presented by the EU's commitment to decarbonize the energy sector for external relations with two countries in the Caspian Sea region – Azerbaijan and Turkmenistan. These relations are particularly embedded in energy relations and EU ambitions to access Caspian natural gas reserves.

Natural gas consumption accounted for about 22 per cent of the EU's final energy consumption in 2012 (Eurostat, 2014). In 2013, the EU consumed 438 billion cubic metres (bcm) of natural gas and it produced 147 bcm domestically (BP, 2014, pp. 22–23). Hence, the EU's reliance on imports of natural gas is very high. As domestic production declines, such import dependence is likely to continue or grow unless the EU moves away from consuming natural gas. Furthermore, EU ambitions to decarbonize by 2050 raise questions about the role natural gas should play in the EU. Unless carbon capture and storage (CCS) technology is rolled out extensively, there is little to no room for continued fossil fuel consumption in a decarbonized EU. What do such realities mean for how relations between the EU and the Caspian region can evolve and develop?

The region includes five countries that border the Caspian Sea – Azerbaijan, Iran, Kazakhstan, Russia and Turkmenistan. Together, these five countries hold reserves of 85 trillion cubic metres of natural gas – about 45 per cent of the world's total proved reserves at the end of 2013 (BP, 2014, p. 20). For the purpose of this chapter, I will focus on the EU's relations with Azerbaijan and Turkmenistan. Azerbaijan is the first port of call for EU access to Caspian natural gas reserves, with Turkmenistan the most likely next access point in the future. The primary motivation for the EU to access natural gas in the Caspian region is to diversify its supply away from Russia (on EU relations with Russia, see Chapter 10). Natural gas production in Kazakhstan is secondary to oil production, and Kazakhstan has major ties to supply Russia with the produced natural gas. As of 2014, broader political relations with Iran do not permit the EU to explore the option of sourcing Iranian natural gas, although this option may open in future. In the short to medium term, natural gas relations with, first, Azerbaijan and, then, Turkmenistan are, or will be, more deeply developed. External relations between the EU and these two countries are based mainly on energy interests of both sides, but the EU aims also to promote good governance and democratic values within these partners. How can decarbonization affect the achievement of these broader aims?

The chapter is structured as follows. First, I describe the role of natural gas in the EU's energy sector today and in a decarbonized future, including EU natural gas production and consumption and infrastructure requirements. Second, I discuss EU–Caspian relations. I highlight the broader political context of the Caspian region, including the interests of several other players in the region. I focus on the natural gas sector and briefly discuss EU relations with Azerbaijan and Turkmenistan. From this discussion, I consider some of the opportunities and challenges that decarbonization might present for EU–Caspian relations, and I highlight the importance of strategic long-term thinking in the EU on the evolution of this relationship. In conclusion, I mention how the decarbonization agenda could help move relations beyond traditional geopolitical considerations of energy security and can present opportunities for heightened emphasis on democratic values.

1. The EU, natural gas and decarbonization

1.1 EU natural gas consumption and production

Natural gas is used in the EU mainly for heating, for some industrial processes, transport and, increasingly, for electricity generation (European Commission, 2010d; IEA, 2011). It is often promoted as the 'cleanest' fossil fuel, with about half the greenhouse gas (GHG) emissions of coal, although there has been some questioning of the 'green credentials' of natural gas (Harvey, 2012; Howarth et al., 2011; Lustgarten, 2011). Its relative abundance globally, with proven reserves in 2013 estimated to last just over 55 years in view of 2013 production rates (BP, 2014, p. 20), and the already in-place technology to exploit and transport it, make gas an attractive option for a short-term transition away from coal. However, before long, natural gas becomes part of the climate problem, as its combustion results in GHG emissions. For decarbonization to be achieved, either natural gas must be eliminated from the energy sector in the medium to long term, or used only in combination with – as yet commercially unviable – CCS technology (Reichardt et al., 2012; see Chapter 1).

The energy security dimension of natural gas supply is particularly important for the EU. The EU as a whole is dependent on a limited number of natural gas suppliers, and on Russia most predominantly. Domestic production of natural gas has been steadily declining in the EU, from 226 bcm in 2003 down to 178 bcm in 2010 and to nearly 147 bcm in 2013 (BP, 2014, p. 22). Over the course of the same decade, natural gas consumption in the EU fluctuated, but has followed a slightly downward trend, especially since 2010 (see Figure 9.1). In 2000, the EU consumed 440 bcm of natural gas (BP, 2011). Highs of 497 bcm and 502 bcm of natural gas consumption were recorded for 2005 and 2010 (BP, 2014). The latest figures for 2013 show the lowest levels of consumption over the previous ten years, at 438 bcm (BP, 2014). While it is difficult

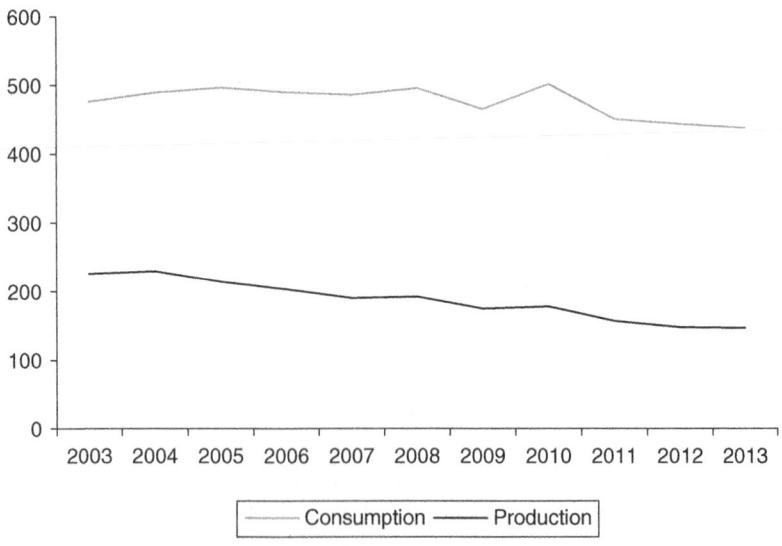

Figure 9.1 EU natural gas consumption and production 2003–2013 (measured in bcm)
Source: Compiled from BP (2014, pp. 22–23).

to describe a trend in EU gas consumption, consumption has not sky-rocketed, as expected following on the rising trends in the 1990s (BP, 2002). However, with the levels of EU production continuing to decline, even steady or slightly declining gas consumption raises concerns about security of supply, at least in the short term.

1.2   EU energy security and natural gas

In addition to declining domestic natural gas production, the EU has experienced two gas supply disruptions that have placed more emphasis on security of supply in natural gas relations. These disruptions took place in 2006 and in 2009, when Russia interrupted supplies to Ukraine, affecting EU member states downstream. The 2006 crisis was short and had no major repercussions on EU member countries as supplies were restored after 36 hours (European Commission, 2008). The ensuing January 2009 crisis was labelled 'unprecedented' (European Commission, 2009). The disruption in gas supplies to Ukraine meant that a number of EU member states were left without adequate supplies of gas for several weeks. These two crises compounded the perception that the EU's reliance on Russia for gas supplies meant it was vulnerable to interruptions in supply (see Keohane & Nye, 1977, for a discussion on 'vulnerability'; see also Chapter 8). The Ukrainian crisis from 2013 onwards further compounds the sense of vulnerability. Strategies to reduce the EU's reliance on natural gas imports from Russia, in particular, are underlined as solutions to free the EU to exert broader influence in external relations with Russia (Erdogdu, 2010; Pozsgai, 2012; Ritter, 2012).

Several responses to increasing dependence on Russia as a single main supplier, continuing import dependence, and perceived vulnerability to supply interruptions may be envisaged. First, the EU could aim to increase domestic production of natural gas. With EU reserves of natural gas expected to last less than 11 years under 2013 production rates (BP, 2014, p. 20), increasing domestic natural gas production is not a feasible option for the EU, unless it invests heavily in unconventional gas (not yet politically feasible in most EU member states). Second, the EU could diversify its sources of natural gas. This particular strategy is one of the main motivations for new natural gas infrastructure connecting the EU to the Caspian. Third, the EU could reduce its consumption of natural gas (by moving to other energy sources or improving energy efficiency). This third option follows a 'decarbonization' logic, as long as the alternative energy sources are non-fossil fuel sources.

Several scenarios and forecasts on decarbonization in the EU suggest that natural gas will play only a limited role in the energy mix in 2050.

Very small amounts may still be required for certain industrial processes, but these would amount to a maximum of about 50 bcm (Heaps et al., 2009). Other scenarios indicate greatly reduced levels of natural gas consumption in the EU by 2050 (EREC, 2010; European Commission, 2011a).[1] With CCS technology advancing more slowly than desired and with limited interest from energy companies in developing or deploying the expensive technology (Odenberger & Johnsson, 2010; Reichardt et al., 2012), decarbonization requires that natural gas consumption in the EU be drastically reduced by 2050 (see also Chapter 1). However, this conclusion is reflected neither in the EU's external relations with Azerbaijan and Turkmenistan nor in the EU's internal decisions on developing new natural gas infrastructure with the region.

1.3   EU natural gas importing infrastructure

Natural gas imports into the EU require infrastructure. Natural gas is most commonly transported via pipeline, but can also be shipped as liquefied natural gas (LNG). Traditional pipeline transportation means that natural gas flows have limited flexibility with regard to both source and destination. The existence of infrastructure linking two markets determines the flow of gas. Therefore, when considering the future role for natural gas in the EU, we must consider the infrastructure in place and any plans for expanding such infrastructure. In this section, I present the capacity of the EU's natural gas importing infrastructure and discuss the necessity for this infrastructure under decarbonization commitments.

While decarbonization means reducing natural gas consumption, the infrastructure in place to import natural gas into the EU continues to expand. Natural gas pipelines and LNG terminals have lifetimes of up to 50 years. Any new infrastructure planned or constructed since 2000, therefore, is likely still to be available for use in 2050 (European Commission, 2010d; European Parliament, 2009).

According to the European Commission's Energy Market Observatory data, EU import pipeline capacity totalled about 441 bcm by 2012. This includes pipelines from Libya, Algeria, Russia and Norway, and transit pipelines through Ukraine and Turkey (European Commission, 2010a; 2010b; 2010c; 2011b; 2011c; 2012). There are also a number of pipeline projects under construction or just constructed since 2011. These include (among others) the Trans-Adriatic pipeline (TAP) in the Southern gas corridor with a capacity of 10 bcm and the second Nord Stream pipeline from Russia to Germany (27.5 bcm). Including these pipelines under construction or just constructed increases pipeline capacity for imports of natural gas into the EU to nearly 479 bcm before

Table 9.1 Natural gas importing infrastructure capacity in EU in 2012 and expected capacity in 2022 (measured in bcm)

	c. 2012	Increases by	c. 2022
Pipeline capacity	441	*38*	479
LNG capacity	181	*78*	259
Total	622	*116*	738

Note: Planned or proposed projects since 2014 not included.
Source: Compiled from European Commission Market Observatory data; Gas LNG Europe (2014); Eurostat (2014).

2020 (a total that does not include planned projects not yet under construction in 2014; see Table 9.1).

Infrastructure for natural gas imports also includes LNG terminals. LNG terminals exist in Belgium, France, Greece, Italy, the Netherlands, Portugal, Spain and the United Kingdom and some small scale LNG in Sweden, with a combined capacity of about 181 bcm per year in 2014 (Gas LNG Europe, 2014). This LNG capacity is planned to increase, with several existing LNG terminals due to expand their capacity, and new terminals under construction in Spain, Italy, France, Lithuania and Poland. Total capacity for LNG imports should therefore increase to about 259 bcm by 2022, again without including planned projects not yet under construction as of late 2014 (Gas LNG Europe, 2014; see Table 9.1).

Total natural gas infrastructure in the EU is therefore *increasing*, and with long operational lifetimes, much of this infrastructure will likely remain in place to 2050, and beyond. By about 2022, natural gas import infrastructure (pipelines and LNG terminals) will reach about 738 bcm, without considering planned projects (see Table 9.1). With natural gas consumption in the EU at about 438 bcm in 2013 (the lowest levels in at least a decade, and including consumption of EU-produced gas), questions must be raised about the *necessity* of such overcapacity in infrastructure, especially when decarbonization objectives to 2050 are taken into account.

Considering the levels of natural gas consumption and production in the EU, the amount of import infrastructure seems excessive, from a decarbonization perspective, but also from the perspective of efficient allocation of economic resources. A large portion of the first round of EU funds under the Connecting Europe Facility (an EU policy package to fund infrastructure projects, agreed in 2013, Regulation (EU) No 1316/2013) have been siphoned into natural gas infrastructure, with smaller portions going to infrastructure projects that aim to progress towards decarbonization (such as smart grid technology).[2] But decarbonization commitments to 2050 point to a further absurdity and risk in the continued expansion of natural gas infrastructure – this expansion represents a real risk of 'carbon lock-in'. The risk is that the very existence of fossil fuel infrastructure may prevent a timely move away from fossil sources of energy and jeopardize the realization of decarbonization aims (Dupont & Oberthür, 2012).

2. The EU, the Caspian and decarbonization

Despite the decarbonization objective of the EU, the Caspian Sea region is considered a key future player for EU energy security for many EU policymakers and politicians, as natural gas reserves in this area are seen as an alternative to Russian gas supplies (Kalyuzhnova, 2005). Azerbaijan is the first of the Caspian Sea countries to agree to provide natural gas to the EU, and Turkmenistan, which has large natural gas reserves, is considered to be a most likely partner for any future supplies of gas from the region. In this section, I place EU relations with Azerbaijan and Turkmenistan in the broader political context in the region. I then discuss the natural gas potential of each of these countries, their governance structure and the institutionalized relations already in existence between the EU and these partners.

2.1　EU–Caspian relations in context

In the case of the Caspian Sea region, several interests intersect, sometimes leaving the EU as an outside actor. From an energy perspective, the United States has long promoted energy relations between the EU and the Caspian Sea region as a way to reduce Russian influence in the region (Ratner et al., 2012, 2013). However, increased energy connection between the EU and Caspian countries has not necessarily led to improved broader relations or democratization (see below).

Russia remains the main player in the region. Much of the pipeline infrastructure in the Caspian runs through Russian territory, and it continues to be a major market for and supplier of gas to the Caspian (Arinc & Elik, 2010). Russia has viewed EU pushes into the region as a threat to its own control over the EU gas market, and to its political influence (Hannibal, 2014). The South Stream pipeline was one of Russia's responses to the perceived problems of interrupted supplies of natural gas to the EU (such as during the 2006 and 2009 gas crises, which Russia argued were due to Ukraine as a transit country, not due to unreliable supply from Russia). It was a pipeline project that aimed to bring Russian natural gas to the EU via the Black Sea, thus bypassing Ukraine. It would have had a capacity of 63 bcm. Intergovernmental agreements between Russia and EU member states Austria, Bulgaria, Croatia, Greece, Hungary and Slovenia were signed to allow the pipeline to be built through these territories, but soured relations between Russia and the EU since the beginning of the Ukraine crisis in 2013 stalled Russian plans for the pipeline. In addition, in a statement on 4 December 2013, the Commission declared that the intergovernmental agreements signed between Russia and several member states were in breach of EU law on the internal energy market and needed to be renegotiated. The rules of 'unbundling' under the EU's energy market liberalization policy mean that infrastructure operators and gas suppliers cannot be the same entity. In the case of South Stream, it is Gazprom that holds these roles. Furthermore, Gazprom cannot be the sole user of the pipeline, according to rules on third party access (see Directive 2009/73/EC concerning common rules for the internal market in natural gas; see Chapter 2). Although construction on Russian territory for the pipeline began, Russia announced in December 2014 that it saw Turkey as a future recipient of natural gas through South Stream instead of the EU.[3]

China has also made moves to access energy resources from the region, with a pipeline to Turkmenistan supplying gas to China since 2009. China's motivations for enhanced relations with the region are based on economic needs, without the democratization ambition of the EU. Thus, energy relations with China can be seen as uncomplicated with the sort of conditionality that may be linked to relations with the EU (Arinc & Elik, 2010). Turkey is also becoming a key player, with ambitions to become an 'energy hub' between the region and the EU (Arinc & Elik, 2010; Marketos, 2009; Sharples, 2013).

These multiple interests are further compounded with the EU's clear strategic interest in moving beyond energy interdependence with Russia, especially in the wake of the Ukraine crisis that began in 2013, its objectives to decarbonize, and its calls for further democratization of society in the region. Russian political influence in the region remains strong, which presents particular challenges for the EU's strategy towards Azerbaijan and Turkmenistan in the context of deteriorating EU–Russia relations. While the Caspian countries may play a balancing game between Russia and the EU, too much enthusiasm for EU ideals may prove damaging for their own relations with their Russian neighbour.

Beyond these broader interests in the region, decarbonization presents opportunities for the EU to move away from energy-based external relations. It can help promote energy independence through more domestically produced renewable energy and heightened levels of energy efficiency (EREC, 2010), rather than continued emphasis on natural gas. However, if the EU continues to pursue a security of supply strategy that promotes diversification of natural gas supply routes away from Russia, both Azerbaijan and Turkmenistan will become increasingly important partners. Turkmenistan holds particular potential as a gas supplier with its large untapped reserves. Considering the democratic and human rights values of the EU, relations with these two countries have been criticized (Boonstra, 2010), even without considering the long-term implications for the achievement of the decarbonization objective.

2.2 Azerbaijan

Natural gas: In June 2013, the Trans-Adriatic Pipeline (TAP) project was selected to transport gas from the Azeri Shah Deniz II gas field to Europe. This announcement concluded years of negotiations and delayed decisions about which project among those proposed in the so-called Southern gas corridor would be selected. The corridor represents a priority infrastructure project for the EU to transport gas from the Caspian Sea region to the EU (through EU policies such as the Trans-European Networks for Energy, the Connecting Europe Facility and the European Energy Programme for Recovery, which supports energy projects in the context of the economic crisis). Other proposed pipeline projects in the corridor included the ill-fated Nabucco pipeline (Erdogdu, 2010; Rowley, 2009).[4] The final decision from the Azeris to favour the smaller TAP pipeline (10 bcm) through Greece and onto Italy came after long discussions, involving not only representatives of the various pipeline consortia, but also members of the EU institutions.[5]

Azerbaijan is a long-time oil producer, but is gradually developing its natural gas production capacity. It has proven gas reserves of nearly 1 trillion cubic metres and has gradually increased its production capacity to reach just over 16 bcm in 2013 (up from 4.6 bcm in 2003) (BP, 2014, p. 22). For the EU, its Shah Deniz II gas field represents a stepping stone for accessing gas in the other Caspian Sea countries (political relations pending) (European Commission, 2010d). The TAP project will be designed to allow for expansion to 20 bcm of capacity in future, if further supplies of gas become available.[6]

Azerigaz, a subsidiary of the state-owned oil company SOCAR, is responsible for natural gas processing, transport, distribution and storage, and Azneft (another SOCAR subsidiary) is responsible for exploration, development and production. While some international investment and joint ventures exist in Azerbaijan, the regulatory framework is generally discouraging for foreign investment in the energy sector, due to corruption and lack of transparency (see below; Alieva, 2014; Ciarreta & Nasirov, 2012; Franke et al., 2010). Azerbaijan's main gas export destinations in 2013 were Russia, Iran and Turkey, with some small amounts of gas reaching Greece through the Turkey-Greece interconnector (EIA, 2013).

Governance: Azerbaijan is governed by an authoritarian presidential regime, with President Ilham Aliyev in power since 2003. Elections take place every five years, with the latest elections in October 2013, and no limit to the number of presidential terms per individual. President Aliyev was re-elected in 2013 with nearly 85 per cent of the vote, in elections that the Organization for Security and Cooperation in Europe (OSCE) reported as not meeting international standards for freedom and fairness. The next elections are scheduled for 2018 (Freedom House, 2014).

Azerbaijan was classified as 'not free' in Freedom House's 2014 'Freedom in the World' publication (Freedom House, 2014, p. 18). Freedom House described regressive trends in freedom in Azerbaijan in 2013, and that the country was locked in a 'downward spiral' that did not present opportunities for the development of liberties in the near future. These downward trends suggest that there is little hope for the 2018 elections to be freer and fairer than those in 2013. Furthermore, Freedom House criticized other democratic governments by arguing that because Azerbaijan is rich in oil and natural gas resources it has 'escaped the condemnation of democratic governments' (p. 7). Azerbaijan's civil liberties were especially damaged in 2013 due to government crack downs on opposition and civil society in the run up to the elections in October.

EU–Azeri relations: EU–Azeri relations are institutionalized in the EU-Azerbaijan Partnership and Cooperation Agreement, which entered into force in 1999. It is a partner country under the European Neighbourhood and Partnership Instrument (ENPI). The ENPI EU-Azerbaijan Action Plan, adopted in 2006, focuses on promoting democratization, human rights, socio-economic reform, poverty alleviation, energy, conflict resolution and further sectoral issues. However, reviews of how the EU is getting on in its promotion of democratic values, civil and human rights in Azerbaijan show a poor record (Alieva, 2014; Franke et al., 2010). Franke et al. (2010) point to the large divergence of interests between the EU and Azerbaijan in their relations. Beyond the energy field, where Azerbaijan views the EU as on the receiving end of relations, Azerbaijan's elites have little interest or incentive to respond to EU efforts at democracy and human rights promotion (Franke et al., 2010).

According to the EU's Azerbaijan country strategy paper for 2007 to 2013 under the ENPI, Azerbaijan is regarded as playing a 'pivotal role' in the EU's energy security of supply (European Union, n.d., p. 5). Azerbaijan is also a part of the Baku Initiative, launched in 2004, that aims to integrate energy markets in the Caspian and Black Sea regions. Internally in the EU, the trans-European network for energy guidelines and the Connecting Europe Facility highlight the priority project status of natural gas interconnections between the EU and the Caspian region through the Southern gas corridor. The energy sphere seems to be the sole EU priority under the ENPI where progress in relations has been made. As Franke et al. (2010) argue, this progress is due to Azerbaijan's own interests in promoting its energy sector and in the EU's perceived dependence (from an Azeri perspective) on receiving supplies of natural gas. Such a perception of relations puts the EU in a tricky situation where, as a receiver of benefits on one element of the relationship, it has little leeway to criticize lack of progress or backsliding on other priorities (such as democratization).

2.3 Turkmenistan

Natural gas: Turkmenistan represents an important potential supplier of natural gas for the EU. Its 2013 confirmed natural gas reserves amounted to 17.5 trillion cubic metres (BP, 2014, p. 20), making Turkmen proven natural gas reserves the fourth largest in the world, after Russia, Iran and Qatar. Turkmenistan has been developing its production capacity over time, and it produced over 62 bcm in 2013 (up from nearly 54 bcm in 2003) (BP, 2014, p. 22). Turkmenistan began to break its large dependency on Russia as its main export market when a pipeline connecting with China was opened in 2009, running through Uzbekistan and Kazakhstan (Boonstra, 2010). Even with gas exports to China, however, Turkmenistan remains locked into certain long-term contracts of gas import and export with Russia, and Russia plays a key transit role for Turkmen gas to other markets. As of 2014, Turkmenistan also exports to Iran, and is planning to connect to the Indian market.

In the wake of a gas crisis with Russia in 2009, Turkmenistan has been looking at several options to diversify its export market, and has made some attempts to open up to foreign direct investment (Arinc & Elik, 2010). The 2009 crisis erupted after an explosion of one pipeline close to the Uzbek border: when the pipeline was fixed, Russia nevertheless refused deliveries through the pipeline. This soured relations between the two countries to an extent and prompted Turkmenistan to reassess options. The EU market is certainly one option for receiving exports of Turkmen gas. At present, however, there are no direct gas infrastructure connections between the EU and Turkmenistan.

Governance: Turkmenistan was ranked as 'not free' by Freedom House in 2014, and received the lowest possible score for political rights and civil liberties (Freedom House, 2014, p. 22). It was listed among the world's 12 most repressive regimes in 2013, along with North Korea and Saudi Arabia, for example. The country is under dictatorship rule and has a very poor record for human rights. After the death of President Saparmurat Nyyazow in December 2006, Gurbanguly Berdimuhammedov was elected president in February 2007, and was reelected in 2012 with 97 per cent of the vote.[7] The OSCE considered that the elections were unfair and lacked freedom. The governmental power resides within the one position of the president.

The Turkmen economy relies on its oil and gas sector. There is little other economic development potential in the country under its current economic set-up, except for some cotton production. Although signs were given that foreign investment in the country would become a possibility, in practice, corruption, poor governance and lack of transparency have made this unrealistic (Arinc & Elik, 2010).

EU–Turkmen relations: EU energy relations with Turkmenistan are institutionalized within an 'Interim Trade Agreement' that was adopted in 1998. The EU has not adopted a negotiated Partnership and Cooperation Agreement with Turkmenistan. The European Parliament has blocked ratification of this agreement due to Turkmenistan's poor human rights record (Boonstra, 2010). A specific memorandum of understanding in the field of energy was agreed in 2008, and Turkmenistan promised 10 bcm of gas to the EU per year (Boonstra, 2010). Turkmenistan is also one of the countries named under the EU's Central Asia Strategy, agreed by the Council in 2007.

Among the priorities for the EU in this relationship is the promotion of economic reform, market reform, education and capacity building, good governance and rule of law, the promotion of civil society, and agriculture and rural development. The EU, however, has few institutionalized opportunities for interaction with Turkmenistan and only promises of energy connections in future. So far, EU pushes for democratic change have had little or no impact on Turkmenistan (Freedom House, 2014).

3. Beyond geopolitics: Opportunities and challenges for EU–Caspian relations to 2050

Decarbonization can present a number of opportunities and challenges for the EU's external relations (see Chapters 1 and 8). It represents a pathway that could move traditional external relations beyond a narrow framing on geopolitics and on control over (fossil) resources, and could open up new avenues for dynamism in the EU's external relations (see Chapter 8). However, this would require EU diplomats and policymakers to take a long-term strategic view of external relations and to place decarbonization at the centre of strategizing and negotiations.

It seems clear that this long-term strategic vision is missing from EU relations with Azerbaijan and Turkmenistan. Short-term security objectives trump any long-term considerations of climate policy objectives or decarbonization. These concerns also trump concerns about human rights and democratic values, especially in relations with Azerbaijan, where mutual interests converge only in the energy sector (Franke et al., 2010). Thus, the potential for change in these relations is hardly recognized, and little high-level political commitment to decarbonization is demonstrated. External relations seem to proceed in a reactionary way, where one crisis leads to a certain short-term strategy, rather than in building long-term strategic visions that can help direct relations down a new path. In this section, I outline a number of opportunities and challenges arising from decarbonization for EU relations with Azerbaijan and Turkmenistan. I look further at the regulatory framework, technology and infrastructure developments in the Caspian region as indicators of potential for the opportunities to be exploited (see Chapter 8).

3.1 Opportunities

First, decarbonization presents an opportunity for greater energy independence in the EU (European Commission, 2011a; WWF, 2007). Among the main strategies to achieve decarbonization by 2050 are a dramatic increase in the share of renewable energy and an increase in energy efficiency (and hence, moving away from fossil fuels and reducing energy consumption) (Heaps et al., 2009). If the EU pursues such an agenda, natural gas will no longer need to play a major role in the EU's energy mix in the long term, and external relations can gradually move away from a framing of EU energy security vulnerability.

Second, decarbonization represents an opportunity to develop relations with partners on new terms. If the EU moves away from a fossil fuel-based economy, its dependence on oil and gas exporting countries (with whom political relations can be strained) may become less relevant. New partners may emerge who see an opportunity for increasing exports of renewable energy in the EU, or new environmentally friendly technologies. The EU could more forcefully promote democracy and human rights values abroad as it opens up to new partners eager to access its large market. Old partnerships may also change. The EU may find it easier to play a stronger hand in its relations with the Russian Federation, Turkmenistan and Azerbaijan, for example. As Connie Hedegaard argued: 'The Ukrainian debate would have been slightly different if we were not that dependent on importing Russian gas'.[8]

Both these opportunities could drastically change EU relations with Azerbaijan and Turkmenistan. The EU could become a more active promoter of human rights and democratic values in the region, with support for rule of law, civil society and education becoming the focus of short to medium-term relations. Long-term relations could focus on the EU and the Caspian region together moving towards decarbonization, as the EU entrenches values of openness, democracy and human rights further into the governance regimes of the region. As Azerbaijan and Turkmenistan become more engaged with the international system, and wish to turn away from too much reliance on Russia, the EU's conditions for improved or closer relations could become more relevant than at present, when mutual interest in energy relations cloud the EU's other objectives in the region (Alieva, 2014; Franke et al., 2010).

Furthermore, as the EU looks for new low-carbon energy partners, Azerbaijan and Turkmenistan may be keen to move beyond their own fossil energy sector. At present, these countries produce little to no renewable energy (except for some hydropower in Azerbaijan) (EIA, 2013). There is, however, potential for considerable wind power, especially along the coastal regions. Until there is an incentive for these countries to develop their potential (such as a large EU market no longer interested in fossil fuels, but in trade in renewable electricity instead), it is unlikely these Caspian countries would move beyond exploiting their fossil fuel resources. Given the reliance of the region on the energy industry for economic development, diversifying this sector into alternative forms of energy presents an opportunity both for economic development in Azerbaijan and Turkmenistan and for the EU's achievement of its decarbonization aims (Arinc & Elik, 2010). Such developments would, however, face a number of transitioning constraints.

3.2   Challenges

The EU's move towards decarbonization also presents some challenges. First, moving to decarbonization presents the risk of reduced 'interdependence'. Relations based on interdependence in the energy sector between the EU and other partners can be exploited as a way of promoting EU values abroad. If the EU moves away from Turkmenistan as an energy partner, for example, the EU may also lose influence in the country. Turkmenistan has increasing energy links with China and Iran – two countries for whom the human rights promotion associated with the EU is irrelevant in external relations (Boonstra, 2010).

Second, and linked to the above point, if the EU wishes to balance relations with Russia and China, it may need to consider carefully the implications of decarbonization. China is a fast-growing renewable technology leader, while also a massive consumer of coal, and relations with the EU are not always straightforward. Russia's interdependent relationship with the EU is based largely on fossil fuel trade. If the EU moves towards decarbonization, will it lose its influence in the region and will Russia gain influence? Or will this allow for stronger EU stances against Russian political manoeuvring in the Caspian region? Will relations with China become more strained as China accesses more energy resources from the Caspian without concern for democratic ideals?

In the cases of Azerbaijan and Turkmenistan, relations cannot yet be described as based on 'interdependence' in the energy realm. For Azerbaijan, perceptions that the EU is a receiver of benefits from Azerbaijan in the energy domain provide little incentive for Azerbaijan to respond to other EU demands (Franke et al., 2010). Relations with Turkmenistan have not yet developed to the extent that considerable energy trade with the EU is taking place. Thus, a move away from an energy interdependence framing of external relations may allow the EU to exert more 'normative power' or 'market power' in its external relations (Damro, 2012; Manners, 2006).

However, the prospect of rising Chinese and continued Russian influence in the region may indeed be of concern to the EU if external relations move beyond geopolitical framings of control over resources. Some signs of interest from Azerbaijan and Turkmenistan to move out from under the shadow of Russian influence are already perceptible (Alieva, 2014; Arinc & Elik, 2010). This could present an opening for the EU to deepen relations beyond the energy sector. In addition, with China aiming to halt its emissions growth by 2030, it remains to be seen if the country will maintain economic relations with the Caspian region to access fossil fuel resources there. For China, at least, relations with Turkmenistan have seemed less political and more based on an economic rationale (Arinc & Elik, 2010). Increased use of natural gas in China instead of coal may be one option for the country to achieve its 2030 emissions goal, which would suggest a greater need for Caspian gas. Chinese strategies to develop domestic renewable energy may temper some of this demand. With these developing (and uncertain) realities, the fear of loss of influence in the Caspian region as compared with Russia and China may be part reality, part exaggeration for the EU.

3.3   Potential for progress: Regulatory framework, technology and infrastructure

In sum, we can say that decarbonization in the EU presents a number of opportunities and challenges for the EU in its external relations with Azerbaijan and Turkmenistan. However, both the opportunities and challenges represent potential for change in relations that, in 2014, are anything but smooth sailing. Concerns over freedom, democratic values and human rights continuously taint relations with these countries, and the EU has so far had little success in promoting its ideals in the region (Alieva, 2014; Barbé & Johansson-Nogués, 2008; Cornell, 2014).

As outlined in Chapter 8, there are a number of concrete factors that can suggest how far decarbonization will influence the development of external relations. The existence of infrastructure compatible with decarbonization pathways, the regulatory framework and the technological development in the partner country can all influence how quickly bilateral relations may advance along lines compatible with decarbonization. For Azerbaijan and Turkmenistan, there is little development on each of these three fronts.

Azerbaijan's infrastructure links to the EU will be through a new natural gas pipeline. More promising infrastructure connections for decarbonization would include electricity grid infrastructure with the EU that would allow for trade in renewable electricity. For Turkmenistan, there are currently no pipeline connections to the EU. However, the technological development for renewable electricity generation in both countries is lacking. Emphasis remains on fossil fuel exploitation as the central foundation of the economy (EIA, 2013). Renewable electricity connections between the EU and these countries may prove unviable due to geographic situations and distance. Nevertheless, the EU could offer assistance in low-carbon technological development as a carrot for future relations. But such an offer may depend strongly on the development of the regulatory framework within both countries. At present, corruption, lack of transparency and poor implementation of the rule of law make Azerbaijan and Turkmenistan unattractive for foreign investment (Freedom House, 2014).

Considering the lack of progress on low-carbon infrastructure connections, technology development, and a transparent regulatory framework in Azerbaijan and Turkmenistan, there is an urgent need and considerable scope for shaping these relations based on a long-term decarbonization strategy. Some first steps in the short to medium term could focus on promoting the openness of society and building up democratic ideals in the region. EU support for the protection of human rights, fundamental freedoms and civil society could be further promoted to move beyond traditional fossil fuel geopolitics towards placing decarbonization at the heart of strategic relations. A gradual development of the regulatory framework towards more openness could then allow for further low-carbon infrastructure connections with the region, along with investment in technology and technology transfer.

It is clear that change within the Caspian region will take time. With entrenched interests in keeping the governance status quo (Franke et al., 2010), EU diplomats and policymakers could focus on the long-term game and the opportunities that EU decarbonization presents for new emphases in EU–Caspian relations.

Conclusions

Considering the EU's commitment to decarbonize its energy sector and move to a low-carbon economy by 2050, the continuing energy security framing of EU–Caspian gas relations may seem surprising. As described above, the EU's natural gas consumption has not risen dramatically over the past decade (although consumption levels have fluctuated). While production of natural gas in the EU continues to fall, the EU's response to this reality has been to promote further natural gas interconnections and diversify its sources of natural gas. The Caspian region represents a source of natural gas with the potential to expand production and export into the future (BP, 2014).

The so-called Southern gas corridor is considered a 'priority' infrastructure project by the EU (European Commission, 2010d). In 2010, agreement was reached between the Shah Deniz II consortium in Azerbaijan and the TAP consortium to supply 10 bcm of gas to Europe per year. The TAP project has the potential to expand its transport capacity to 20 bcm in future if required. This is just one of several projects that are planned or under construction for expanding the EU's capacity to import natural gas. With lifetimes of about 50 years, this infrastructure would be in place beyond 2050, leading to risks of carbon lock-in to fossil fuel infrastructure in the EU.

This situation points to several conclusions. It seems that the long-term commitment to decarbonize by 2050 is not taken seriously by EU diplomats and energy policymakers. In this case, emphasis on new infrastructure and increasing supplies of natural gas seems counter to decarbonization objectives – natural gas will become part of the climate problem before too long. As such, external relations with the Caspian region seem to continue on a business-as-usual basis, without strategic long-term planning of the consequences and implications of EU decarbonization.

Considering the persistent short-term view EU–Caspian relations follow, the opportunities presented by EU decarbonization remain unexploited and the challenges unaddressed. Significant opportunities to move these relations beyond traditional geopolitical considerations of control over fossil resources exist, as well as challenges that link to the relative influence of the EU in comparison to Russia and China in the region. Thinking strategically in the long term can help exploit the opportunities and address the challenges so as to develop fruitfully the EU's external relations with the region.

Relations based on decarbonization will develop only slowly between the EU and the Caspian, considering that low-carbon infrastructure connections, technology and the openness of the regulatory framework in Azerbaijan and Turkmenistan are neither developed nor prioritized. The EU may need to play a long-term game, but doing so could provide the opportunity for stronger support for democratization and promotion of human rights in the region. Without strategic long-term thinking about how external relations with the Caspian region could and should evolve under decarbonization, the EU may miss potential opportunities to diversify or reprioritize relations.

Notes

1. One exception is Eurogas, which argues for long-term continued use of gas – in combination with CCS technology (Eurogas, 2011).
2. See Euractiv report, 23 September 2014: 'EU risks wasting billions on gas infrastructure "white elephants"', http://www.euractiv.com/sections/energy/eu-risks-wasting-billions-gas-infrastructure-white-elephants-308625, date accessed 25 September 2014.

3. See Euractiv report, 2 December 2014 'Russia says South Stream project is over', date accessed 3 December 2014.
4. The Nabucco pipeline project did not receive the backing of Azerbaijan for supplies of gas to Europe. It was an EU-backed proposed project to build a pipeline between the Caspian and Austria.
5. For example, in January 2011, then European Commission President, Jose Manuel Barroso, promised visa facilitation for Azeri nationals in exchange for promises of supplies of natural gas to the European Union (see http://www.euractiv.com/energy/barroso-tops-azeri-gas-deal-visa-news-501255, date accessed 4 March 2013).
6. See www.trans-adriatic-pipeline.com for more detail, date accessed 22 July 2013.
7. See https://www.cia.gov/library/publications/the-world-factbook/geos/tx.html, date accessed 28 August 2013.
8. See Euractiv report, 10 March 2014 'Hedegaard: "Ukrainian crisis shows we need to reduce our energy dependency"', http://www.euractiv.com/video/hedegaard-ukrainian-crisis-shows-534022, date accessed 13 March 2014.

References

Alieva, L. (2014) 'Azerbaijan and the Impact of the Lack of Democratisation on Relations with the EU', *European View, 13*(1), 39–48.

Arinc, I. & Elik, S. (2010) 'Turkmenistan and Azerbaijan in European Gas Supply Security', *Insight Turkey, 12*(3), 169–190.

Barbé, E. & Johansson-Nogués, E. (2008) 'The EU as a Modest "Force for Good": The European Neighbourhood Policy', *International Affairs, 84*(1), 81–96.

Boonstra, J. (2010) 'The EU-Turkmenistan Energy Relationship: Difficulty or Opportunity?', *FRIDE Policy Brief, 5*(October 2010).

BP (2002) *BP Statistical Review of World Energy June 2002* (London: British Petroleum).

BP (2011) *BP Statistical Review of World Energy June 2011* (London: British Petroleum).

BP (2014) *BP Statistical Review of World Energy June 2014* (London: British Petroleum).

Ciarreta, A. & Nasirov, S. (2012) 'Development Trends in the Azerbaijan Oil and Gas Sector: Achievements and Challenges', *Energy Policy, 40*, 282–292.

Cornell, S. E. (2014) 'Underestimating Yourself: The EU and the Political Realities of the Eastern Neighbourhood', *European View, 13*(1), 115–123.

Damro, C. (2012) 'Market Power Europe', *Journal of European Public Policy, 19*(5), 682–699.

Dupont, C. & Oberthür, S. (2012) 'Insufficient Climate Policy Integration in EU Energy Policy: The Importance of the Long-Term Perspective', *Journal of Contemporary European Research, 8*(2), 228–247.

EIA (2013) *Caspian Sea Region* – diverse country profiles from the US Energy Information Administration, http://www.eia.gov/countries, date accessed 27 October 2014.

Erdogdu, E. (2010) 'Bypassing Russia: Nabucco Project and Its Implications for European Gas Security', *Renewable and Sustainable Energy Reviews, 14*, 2936–2945.

EREC (2010) *RE-Thinking 2050: A 100% Renewable Energy Vision for the European Union* (Brussels: European Renewable Energy Council).

Eurogas (2011) *Eurogas Roadmap 2050* (Brussels: Eurogas).

European Commission (2008) 'Communication... Concerning Measures to Safeguard Security of Natural Gas Supply', COM(2008) 769.

European Commission (2009) 'Proposal for a Regulation of the European Parliament and of the Council Concerning Measures to Safeguard Security of Gas Supply and Repealing Directive 2004/67/EC', COM(2009) 363.

European Commission (2010a) *Country File: Libya* (Brussels: Market Observatory for Energy).

European Commission (2010b) *Country File: Russia* (Brussels: Market Observatory for Energy).

European Commission (2010c) *Country File: Ukraine* (Brussels: Market Observatory for Energy).

European Commission (2010d) *Energy Infrastructure Priorities for 2020 and Beyond: A Blueprint for an Integrated European Energy Network* (Luxembourg: Publications Office of the European Union).

European Commission (2011a) 'Energy Roadmap 2050', COM(2011) 885/2.

European Commission (2011b) *Country File: Algeria* (Brussels: Market Observatory for Energy).

European Commission (2011c) *Country File: Norway* (Brussels: Market Observatory for Energy).

European Commission (2012) *Country File: Turkey* (Brussels: Market Observatory for Energy).

European Parliament (2009) *Directorate-General for Internal Policies. Policy Department A: Economic and Scientific Policy, 'Gas and Oil Pipelines in Europe'* (Brussels: European Parliament).

European Union (n.d.) *European Neighbourhood Partnership Instrument: Azerbaijan*, Country Strategy Paper, 2007–2013 (Brussels: European External Action Service).

Eurostat (2014) 'Statistics Online Database', http://epp.eurostat.ec.europa.eu/portal/page/portal/eurostat/home/, date accessed 5 May 2014.

Franke, A., Gawrich, A., Melnykovska, I. & Schweickert, R. (2010) 'The European Union's Relations with Ukraine and Azerbaijan', *Post-Soviet Affairs, 26*(2), 149–183.

Freedom House (2014) *Freedom in the World 2014. The Democratic Leadership Gap* (Washington, D.C.: Freedom House).

Gas LNG Europe (2014) *LNG Map*, (Brussels: Gas Infrastructure Europe).

Hannibal, M. (2014) 'The Power of Energy Politics', *European View, 13*(1), 73–78.

Harvey, F. (2012) 'Gas Rebranded as Green Energy by EU', *The Guardian*, 29 May 2012, http://www.theguardian.com/environment/2012/may/29/gas-rebranded-green-energy-eu, date accessed 4 June 2012.

Heaps, C., Erickson, P., Kartha, S. & Kemp-Benedict, E. (2009) *Europe's Share of the Climate Challenge: Domestic Actions and International Obligations to Protect the Planet* (Stockholm: Stockholm Environment Institute).

Howarth, R. W., Santoro, R. & Ingraffea, A. (2011) 'Methane and the Greenhouse-Gas Footprint of Natural Gas from Shale Formations: A Letter', *Climatic Change, 106*(4), 679–690.

IEA (2011) *World Energy Outlook 2011: Are We Entering a Golden Age of Gas?* (Paris: International Energy Agency/OECD).

Kalyuzhnova, Y. (2005) 'The EU and the Caspian Sea Region: An Energy Partnership?' *Economic Systems*, 29, 59–76.

Keohane, R. O. & Nye, J. (1977) *Power and Independence: World Politics in Transition* (Boston, MA: Little Brown).

Lustgarten, A. (2011) 'Climate Benefits of Natural Gas May Be Overstated', *Scientific American*, http://www.scientificamerican.com/article.cfm?id=climate-benefits-natural-gas-overstated, date accessed 8 March 2012.

Manners, I. (2006) 'European Union "Normative Power" and the Security Challenge', *European Security*, 15(4), 405–421.

Marketos, T. (2009) 'Eastern Caspian Sea Energy Geopolitics: A Litmus Test for US-Russia-China Struggle for the Geostrategic Control of Eurasia', *Caucasian Review of International Affairs*, 3(1), 2–19.

Odenberger, M. & Johnsson, F. (2010) 'Pathways for the European Electricity Supply System to 2050: The Role of CCS to Meet Stringent CO_2 Reduction Targets', *International Journal of Greenhouse Gas Control*, 4(2), 327–340.

Pozsgai, P. (2012) 'The Evolution of EU Security of Gas Supply Policy', *EDI Quarterly*, 4(3), 35–37.

Ratner, M., Belkin, P., Nichol, J. & Woehrel, S. (2012) 'Opportunities and Obstacles for European Alternatives to Russian Natural Gas', *EDI Quarterly*, 4(3), 42–44.

Ratner, M., Belkin, P., Nichol, J. & Woehrel, S. (2013) *Europe's Energy Security: Options and Challenges to Natural Gas Supply Diversification* (Washington, D.C.: Congressional Research Service).

Reichardt, K., Pfluger, B., Schleich, J. & Marth, H. (2012) 'With or Without CCS? Decarbonising the EU Power Sector', *Responses Policy Update*, 3(July 2012).

Ritter, A. (2012) 'Managing the EU's Gas Security of Supply: Not Without Ukraine', *EDI Quarterly*, 4(3), 40–41.

Rowley, M. (2009) 'The Nabucco Pipeline Project – Gas Bridge to Europe?' *Pipeline and Gas Journal*, 236(9), 72–73.

Sharples, J. D. (2013) 'Russian Approaches to Energy Security and Climate Change: Russian Gas Exports to the EU', *Environmental Politics*, 22, 683–700.

WWF (2007) *Climate Solutions. WWF's Vision for 2050* (Gland, Switzerland: WWF International).

10
Evolutions and Revolutions in EU–Russia Energy Relations

Olga Khrushcheva and Tomas Maltby

Introduction

Historically, EU–Russia energy trade has focused on hydrocarbons, and despite the high levels of interdependence, relations have become increasingly politicized and, arguably, securitized in the 2000s. This is due to the disruptions of Russian gas supplies in 2006 and 2009 and the Ukraine crisis in 2013/2014 contributing to concerns about security of Russian energy supplies and EU demand (specifically regarding gas) and the interdependence of the energy trade between the EU and Russia.

In this chapter, we evaluate EU–Russia cooperation on energy and decarbonization. While the EU has more ambitious decarbonization targets than Russia, improving energy efficiency and the integration of renewable energy sources (RES) is also a stated Russian priority. Careful analysis of the prospects for cooperation demonstrates the potential mutual benefits for both the EU and Russia. These benefits include (among others): opening the Russian market for EU technology transfer and investment; ensuring security of energy supply for both the EU and Russia; and security of demand for Russia by creating opportunities for RES export.

We argue that while the potential for much closer cooperation on energy is significant and offers economic, environmental and potential broader geopolitical benefits for the EU and Russia, there are significant obstacles to the realization of these opportunities. This chapter suggests possible ways to overcome these challenges. However, the broader aspects of EU–Russia (energy) relations need to be considered, including the need for greater institutionalization and legislation.

This chapter is divided as follows: we discuss, first, the general background of EU–Russia energy relations; second, the institutionalization of cooperation on decarbonization; third, barriers to achieving decarbonization goals and opportunities for increasing energy efficiency and the share of RES; and, finally, we reflect upon the implications of the 2013/2014 Ukrainian crisis for the development of the energy and decarbonization relationship between the EU and Russia.

1. Background

There has long been a strong mutual interdependence between Russia as an energy exporter and the EU as an energy importer. In 2010, Russia provided 35 per cent of EU gas imports, 32 per cent of oil imports and 27 per cent of coal imports (Eurostat, 2013). In 2013, Russia was the EU's third most important trading partner, and the EU's trade deficit with Russia increased from 30 billion EUR in 2002 to 90 billion EUR in 2012, primarily due to energy imports and increasing energy prices (Eurostat, 2014). In 2014, Russia was the single largest supplier of natural gas to the EU (European Commission, 2014b, p. 44). Potential to replace Russian gas is limited by existing long-term energy contracts, and the cost of diversification (European Commission, 2014b, p. 52). Therefore, in the medium term (to about 2030) the conventional energy trade relationship may continue, though this could be affected by pressures to diversify the EU's energy supplies as a result of deteriorating political relations in 2014. Conventional energy trade could decrease markedly in the long term with the successful implementation of the EU's decarbonization policy objectives.[1]

Significant potential for mutual trade benefits associated with decarbonization could develop into a supplementary energy trade relationship, through technology export and investment from the EU and import from Russia of RES electricity. However, there are a number of obstacles to this potential being realized, including the strained political relations.

Since the 1970s, EU–Russia energy relations have been limited to trade in hydrocarbons. The Soviet Union, and later Russia, was considered a reliable energy supplier until the early 2000s. Since then, EU–Russia energy relations have taken a turn towards securitization (Khrushcheva, 2011). EU leaders are concerned with the slow progress of energy sector liberalization in Russia, and high levels of EU dependence on Russian gas supplies (Stulberg, 2011, p. 2). After the 2006 and 2009 disruptions of Russian gas supplies to the EU (of 3 and 20 days respectively), many commentators focused on the security element of EU–Russia relations (including Aalto, 2008; Youngs, 2009).

Kirchner and Berk argued that 'prospects that energy security will contribute in the foreseeable future to cooperation between Europe and Russia are low' (2010, p. 877), particularly as a result of divergent understandings among member states of energy security and the reliability of Russia as a supplier. With regard to EU–Russia decarbonization strategies, the pattern is similar. In 2004, Laroui et al. were optimistic about the outlook for joint implementation of renewables projects and investment in Russia, within the context of the Kyoto Protocol (Laroui et al., 2004). However, such hopes were wide of the mark, and Russia decided not to participate in the Kyoto Protocol post-2012. In 2013, Youngs noted the unrealized potential of the EU's decarbonization strategy to 'overcome the zero-sum dynamics of gas pipeline politics' and improve EU relations with Russia (p. 430). However, we argue that the EU's decarbonization objectives offer an opportunity to develop a mutually beneficial relationship between the two parties that will have political, economic and environmental benefits if considerable obstacles are overcome. Political will from both parties is required as broader EU–Russia relations in 2014 moved the political focus away from decarbonization. The section below provides a brief overview of the potential benefits for both the EU and Russia.

2. Energy decarbonization relations: Opportunities and barriers

The EU has struggled to export its energy market liberalization objectives or to develop an institutionalized EU–Russia energy relationship. An important development was the 'Roadmap of the EU–Russia Energy Cooperation until 2050' as part of the ongoing (since 2000) EU–Russia Energy Dialogue. The Russian Ministry of Energy and the European Commission's DG Energy were responsible for the negotiations, which led to a proposal in February 2011 and publication of the roadmap in March 2013 (European Commission & Russia, 2011; 2013). The roadmap analyses the impact on EU–Russia energy relations of different scenarios, including the increased use of RES. The objective is to consider the long-term perspective of mutual energy relations, providing recommendations for steps to enhance EU–Russia energy cooperation, including in energy efficiency and RES. These new developments correspond to commitments made by Russia and the EU in international forums on tackling climate change (including the Kyoto Protocol), and had strategic importance for energy securities of supply and demand.

The EU and Russia already had a history of cooperation on alternative energy. The flexible mechanisms of the Kyoto Protocol aimed to encourage investment in emission reduction projects in industrialized or developing countries (Kyoto Protocol, Articles 6 and 12). Fankhauser and Lavric named Russia as one of the countries with 'the highest scope for, and the lowest cost of producing, emission reductions' (2003, p. 7). Russia is host to 208 out of 761 projects in industrialized countries (UNEP, 2014). These projects aim to achieve emission reduction in a flexible and cost-efficient manner by using foreign investment and technology transfer to earn emission reduction units for one country from a project in another.

Similarly, the Commission argues that third countries, including Russia, could benefit from an 'exchange of best practice on support scheme reform' and that there is the possibility to 'facilitate international cooperation on renewable energy development by enabling full use of the cooperation mechanisms... [to] enable Europe to import additional renewable electricity' (European Commission, 2012a, p. 13). To achieve a 20 per cent target for the share of renewable energy in final energy consumption in the EU by 2020, the EU's Renewable Energy Directive (Directive 2009/28/EC) includes mandatory national targets and allows 'joint projects with one or more third countries' to contribute to these.

Cooperation with third countries is thus permitted under EU internal RES legislation, and there may be mutual benefits for both Russia and the EU to drive down costs and exploit growing markets for technology exports. The sparsely populated North-West of Russia holds great potential for cost-efficient onshore wind and biomass, in particular (European Commission & Russia, 2013; IFC, 2011). Such projects may produce less expensive RES electricity than imports from North Africa, which would require costly submarine infrastructure (Boute & Willems, 2012, p. 625). These projects could build on existing limited cooperation in RES trade, where North-West Russian wood pellets are imported into the EU. Further cooperation could lead to improved relations through a field of energy that is less politicized than that of gas supplies, where disruptions of Russian gas supplies to Europe have raised concerns within the EU about Russia using energy as a political tool (Handke & de Jong, 2007; Maltby, 2013). Such projects would also fit within the EU's definition of energy security, that of providing 'safe, secure, sustainable and affordable energy' (European Commission, 2011b, p. 4). The EU–Russia Energy Roadmap to 2050 also underlines that both Russia and the EU are interested in 'projects that could lead to the export of electricity produced from renewable resources from Russia to the EU' (European Commission & Russia, 2013, p. 23).

While the Commission highlights that the internal energy market 'should also facilitate market entry and integration for new players' (European Commission, 2012a, p. 7), in 2012 the Commission's plans for the development of RES in and with neighbouring countries did not make reference to any detailed plans with Russia (European Commission, 2012a). However, the discussion on post-2020 EU targets on RES and GHG emission reduction opens the future of RES projects with third countries (European Commission, 2014a), although financing for such projects remains uncertain (see below).

A 'shift of EU–Russia energy relations from a pure supplier–consumer relationship towards a more technology-based cooperation [with] significant joint cooperation' is part of the aim of the creation of a Pan-European Energy Space by 2050 (European Commission & Russia, 2013, pp. 5–6; see also Chapter 8). However, the legal framework is lacking. Efforts to promote an open and competitive market under the Energy Charter Treaty[2] resulted in Russia announcing its withdrawal from the treaty in August 2009. Overall, the difficulties in exporting energy market liberalization beyond the EU remain a major problem in the further institutionalization of EU–Russia energy relations.

2.1 Renewable energy: An opportunity

Future Russian energy trends indicate far more modest aims than the EU in terms of decarbonization. In EU decarbonization scenarios, RES reaches at least 55 per cent in gross final energy consumption in 2050 (see Chapters 1 and 3), while the Russian energy sector will continue to rely heavily on traditional energy resources, with domestic consumption of natural gas increasing (Overland & Kjaernet, 2009, p. 26). The use of RES is underdeveloped in Russia despite almost every region in Russia demonstrating significant potential (Tynkkynen & Aalto, 2012, p. 98).

There are examples of small-scale RES projects in Russia, including hydropower stations exporting electricity to Finland (IFC, 2008). From the 1970s, Russia developed industrial production of energy from biomass, which it began to export in limited volumes in 2003 (Tynkkynen & Aalto, 2012, p. 102). In 2009, Russia produced around a million tonnes of biomass, two thirds of which were exported to the EU (Kulikova, 2010). The Commission noted the potential for expanding imports of biomass from Russia in its 2050 Energy Roadmap (European Commission, 2011a, p. 11). In 2014, the US Department of Agriculture estimated that Russia had the potential to increase biofuel production tenfold. However, renewables including biofuels represented an insignificant share of Russia's energy mix, estimated at 1.2 per cent, with biomass accounting for 0.5 per cent (US Department of Agriculture, 2014). Also, EU–Russia biomass trade creates a specific environmental problem – the transportation of biomass to the EU contributes to CO_2 emissions. To avoid this, some suggest that Russia should rather export electricity produced from biofuels[3] (Boute & Willems, 2012, p. 624).

The Russian Government aims to achieve 'a leading position in the development of RES' by 2020 (Russia, 2008, p. 15). The Russian energy strategy to 2030 (Russia, 2010) mentions the importance of developing RES for both environmental security and energy security. Domestic energy consumption in Russia is increasing, especially the consumption of natural gas.[4] Furthermore, the supergiant natural gas fields operational in 2013 are diminishing. The development of new oil and gas fields requires significant investment, raising concerns about Russia's ability to meet future international demand (Stern, 2009). Increasing energy efficiency and RES may resolve the potential problem with domestic and export supplies, since 'the technical potential ... exceed[s] the current energy consumption in the country more than fourfold' (Tynkkynen & Aalto, 2012, p. 98).

Despite this potential and a 2009 Russian government objective to increase RES production to 4.5 per cent by 2020 (compared to the 20 per cent target for the EU) (Russia, 2010), the interim 2010 target of 1.5 per cent was missed and the share of RES in Russian energy consumption remained under 1 per cent (IFC, 2011, p. 7). This raises questions about the commitment of the Russian government to increasing RES deployment.

The Russian government foresees that RES growth will need to be facilitated by state support and international cooperation (European Commission & Russia, 2013, p. 21). To enhance EU–Russia cooperation on RES, the EU is in a position to share expertise in developing RES, and can act as an investment partner. Cooperation could be expanded to technology partnerships. The EU's Strategic Energy Technology (SET) plan and Horizon 2020 research programme are the EU's main contribution to driving research and development in key energy technologies (European Commission, 2012a, p. 10). The EU–Russia roadmap underlines the importance of collaboration on energy research (European Commission & Russia, 2013, p. 4).

Financing RES is a major challenge. Russia's Energy Strategy to 2030 predicts that the total capital investment required is 355–554 billion USD (Russia, 2010), but the government is prepared to finance only 8 per cent of this. Russia's lack of a comprehensive and coherent legislative framework results in limited success in attracting private investment (see below). The EU promoted the use of RES in Russia, through the Technical Assistance to the Commonwealth of Independent States programme (2000–2006) and a 2007–2009 project focusing on knowledge exchange and increasing public awareness of RES (European Commission, 2009). Individual member states (Germany, France, Denmark, Finland and Italy) were also involved in projects on energy efficiency, electricity and RES in Russia in 2013 (RuDanEnergo, 2013; EDF, 2013). The effect of these small projects was limited, beyond indicating a willingness in Russia to receive funding and expertise, which may at best pave the way for further and larger-scale joint projects between the EU and Russia.

2.2   Energy efficiency: An opportunity

Although there has been relatively slow improvement in Russia compared to most former Soviet Republics (IFC, 2008, p. 7), energy intensity (the amount of energy necessary to generate one unit of GDP) in Russia decreased by a third between 2000 and 2012 (Enerdata, 2013). However, Russian domestic energy consumption remains highly energy intensive and inefficient. The Russian 2030 energy strategy 'finds considerable untapped potential in organizational and technological energy saving', which could reduce domestic energy consumption by 40 per cent by 2030 compared to 2005 (Russia, 2009; 2010, pp. 30, 138). This is potentially a cost-effective alternative to increasing energy supplies (IFC, 2008, p. 6), and collaboration on energy efficiency could provide a large market for industrial energy-efficient products exported from the EU, while contributing to energy security.

Within the EU–Russia Energy Dialogue, an Energy Efficiency Thematic Group provides a framework for enhanced cooperation in the context of the EU's decarbonization objectives. In 2010, the Group became a key part of the EU–Russia Partnership for Modernization. In December 2012, a Memorandum of Understanding was agreed on a number of projects, including 200 million EUR of European financing (European Investment Bank and European Bank for Reconstruction and Development) for a new gas-fired combined heat power plant (EU–Russia Partnership for Modernization, 2012), and long-term lending commitments for infrastructure and energy efficiency investments (EEAS, 2012). However, in October 2013, the Commission's list of 248 key energy infrastructure 'projects of common interest' included no collaborative projects involving Russia (European Commission, 2013).

There are also obstacles to the implementation of energy efficiency measures in Russia. In 2009, Russia adopted a law on energy saving and energy efficiency increase, but this is only applicable to 12 per cent of electricity consumption (Russia, 2009), and has been poorly implemented. An International Finance Corporation (IFC) report at the time claimed that '80 per cent of energy efficiency potential is financially attractive, but [...] current federal and regional legislation on energy efficiency is largely declarative and does not address key barriers such as lack of information and insufficient access to long-term funding' (IFC, 2008, p. 6). Energy efficiency does, however, provide an opportunity for best practice information sharing with the EU. Gusev (2013, p. 5) argues that the economic imperative, and possible legislative requirement, to improve energy efficiency within Russia will increase demand for EU technologies. Major federal companies responsible for electricity generation and transmission are already largely reliant on foreign technology. There is then a mutual benefit for both the EU and Russia in the short term, though a maturing RES and energy efficiency industry in Russia could provide competition for EU firms. Energy savings are likely to become more economically attractive for industry to maintain export competitiveness. Electricity prices in Russia are predicted to reach EU levels in 2015–2016 (from almost half in 2010) due to rising gas prices, the withdrawal of state subsidies, and the requirement for distribution grid investment (IEA, 2011, p. 600). Gusev considers the planned increase in electricity transmission fees in Russia a catalyst for a greater proportion of decentralized generation (with correspondingly lower transmission costs), which would favour the development of RES (2013, pp. 1–4). Rising conventional energy prices in Russia are likely also to stimulate investment in energy efficiency measures. This could occur regardless of government commitment to decarbonization.

2.3 Barriers to cooperation: Framework for investment

Regulatory instability and unpredictability is one key concern for foreign investors in RES. They are dependent on limited and frequently changing state support for the financial viability of their investments, and the Russian investment climate has been characterized as unpredictable and unstable for foreign investment in strategic industries (IFC, 2011, pp. 51, 68). For example, BP's Russian joint oil venture TNK-BP was subject to 'unprecedented investigations, proceedings, inquiries and other burdens' (BP chief executive Dudley cited in Belton, 2008). Foreign companies wishing to obtain a controlling stake in a company operating in a strategic sector, or to buy more than 10 per cent of larger oil and gas deposits, need to obtain the approval of a governmental commission (Pleines, 2009, p. 74). Legislation outlines that Russian companies have priority in signing agreements, that 80 per cent of the personnel should be Russian, and that the investor pays the State, either in shares of the resources extracted or in shares of the product sales (Maican, 2009, p. 11). The 2013 amendments to the Russian renewable and energy efficiency policy included a local content policy that requires project developers to use national technology. For example, wind and solar installations must be at least 65 and 70 per cent, respectively, assembled or produced in Russia from 2016 onwards (Russia, 2013b). Due to the incoherence and gaps in Russian legislation on FDI, Cameron (2011) mentioned in an interview that small and medium sized foreign companies are reluctant to invest to Russian projects.

With regard to RES and energy efficiency, the main problem is not so much the perceived economic and political sensitivity of the sector, but the lack of a 'functioning regulatory framework at the federal level to make investments in renewable energy commercially viable' (IFC, 2011, p. 3). For instance, the Russian-German Energy Agency, set up in 2008 to support the development of energy efficiency projects in Russia, closed down in 2013 due to 'the lack of interest and investments from the Russian side, and difficult investment conditions in Russia' (Makarychev & Meiser, 2014, p. 7).

In August 2012, Russia joined the WTO – a development that then EU Trade Commissioner Karel De Gucht claimed would 'facilitate investment through a more predictable legislative framework [and] have a gradual positive impact on Russian business as it will improve investment conditions and competition on the Russian market' (cited in European Commission, 2012b). WTO membership will likely put pressure on Russia to follow the legal principles of international trade providing insurance to potential investors.

2.4 Barriers to cooperation: Energy companies

Under the Copenhagen Accord of 2009, Russia committed to a 15–25 per cent GHG emission decrease by 2020, compared to 1990. However, the Accord was not legally binding and Russia did not commit to further reductions under the Kyoto Protocol's second commitment period, from 2013. The 2009 Russian Energy Strategy to 2030 set a modest objective to limit (but not decrease) emissions in 2030 to 100–105 per cent of their 1990 levels (Russia, 2010, p. 129). Despite this, a September 2013 Presidential Decree (no. 752) committed Russia to a 25 per cent reduction in GHG emissions by 2020 (relative to 1990 levels) and set a target of 25–30 per cent by 2030 (Russia, 2013c). This indicates a renewed commitment to tackling climate change, if targets are enforced. However, the role of hydrocarbons in the economy and the objectives of Russian energy companies is an obstacle to achieving these objectives. In Russia's 2030 strategy, domestic natural gas consumption in Russia's energy mix is predicted to decline from 54 per cent in 2008 to 47 per cent in 2030, and oil is expected to decline from 24 per cent to 20 per cent (Russia, 2010, p. 59).

Bazarova (2013) claims that in most major oil and gas producing states in 2013 the powerful interests of the energy industry clashed with the potential benefits of increased investment in low-carbon energy sources. The major energy companies have acknowledged the importance of technological developments in alternative energy since the early 2000s, and a Gazprom representative claimed in 2010 that the company was committed to alternative energy sources (Gazprom interview, 2010). However, in 2010 Gazprom did not prioritize major financial support for RES in its investment policy. Russian experts confirm that, in the early 2010s, the Russian government and major energy companies lacked commitment to developing RES (Simonov, 2010; Gazprom interview, 2010; Pichkov, 2010). In 2013, the major energy companies are the main investors in Russian RES, but 'the traditional energy sources play too important [a] role in the Russian economy to popularize alternative energy' (Bazarova, 2013).

A review of annual reports of major energy companies illustrates their rather limited commitment to RES. Gazprom, together with two other Russian companies and one Italian company, aimed at the production and potential export of biogas from biomass to the EU using existing infrastructure (Gazprom, 2012, p. 97). Despite being a more controversial source of energy, the development of biogas corresponds with the goals of the EU to reduce GHG emissions. The 2009 Renewable Energy Directive highlighted that greater integration of biofuels provides 'significant environmental advantages in terms of heat and power production' (Directive 2009/28/EC, paragraph 12). In 2012, LUKoil bought the Romanian company Land Power in order to construct wind turbines, with a targeted output of 200,000 MW/h per year (LUKoil, 2013), and shares in two Bulgarian RES companies (Kavarna & Long Man) (LUKoil, 2012, p. 40). However, such investments remain rather insignificant fringe activities of the major energy companies. The integration of RES into Russian energy company strategies and production mix is thus still in the early stages.

2.5 Barriers to cooperation: Civil society and public opinion

Opinion polls conducted in Russia in 2013 demonstrate that the Russian population recognizes the importance of RES, with 52 per cent stating it is necessary to amend the existing energy strategy to integrate alternative energy sources (Energy Policy Russia, 2013a). However, 48 per cent of respondents opposed, and only 15 per cent supported, the increase of domestic energy tariffs to promote the integration of alternative energy (Energy Policy Russia, 2013b). These figures indicate that public opinion in Russia may not be strongly in favour of decarbonization, which may represent a barrier to stimulating EU–Russia decarbonized energy relations.

Makarova claims that 'NGOs struggle with unstable financial support, little experience in defending their interests, and insufficient professional capacity and expertise' (cited in Sharmina et al., 2013, p. 380). Sharmina et al. conclude that a 'distrust in "authorities" and the weakness of civil society are a handicap when it comes to long-term national priorities', and that Russia's environmental agency, RosHydromet, is a 'passive observer' (2013, p. 389). There is evidence that public awareness and concern about climate change are lower in Russia than in the EU, though they have increased since the 1990s (Sharmina et al., 2013, p. 382). The continuing subsidies for fossil fuel generation are an obstacle to public acceptance of decarbonization policies and their financing. Such subsidies (amounting to 2.7 per cent of Russia's GDP in 2010) also pose an impediment to investment by energy companies in Russia (IEA, 2011). The political sensitivity of price increases in the energy sector is considered a major barrier to support of Russian RES (IFC, 2013, p. 4). It will require further effort on behalf of the Russian government to promote the benefits of combatting climate change, and address the related (short-term) costs of developing a Russian RES market.

In 2013 the Russian government introduced the Renewable Energy Source Development Measures with a capacity-based RES support scheme (Russia, 2013a). The government designed this regulatory framework to provide a revenue stream, and increased commercial viability, for RES while limiting the amount of capacity supported and the operating costs of projects. This is in part to limit energy price increases by limiting subsidies.

2.6 Barriers to cooperation: Limits to institutionalized relations

At the 2000 EU–Russia Summit, energy was highlighted as the area with 'most potential to lead the European subcontinent into deeper, mutually beneficial integration' (Council of the European Union, 2000).

The EU–Russia Energy Dialogue was initiated the same year. Aalto (2008, p. 37) argued that one strength of this forum has been that it has avoided the significant politicization that has been found elsewhere with regard to energy. Within the Dialogue, the European Commission functions as a 'facilitator', creating an institutional and a political framework for both industry and politicians to meet and debate. At the same time, the forum has been accused of being 'notable mainly for its lack of substantial policy output' (Hadfield, 2008, p. 234), with 'policy-making in the dialogue ... essentially technical ... [with a] narrow remit' (Aalto & Korkmaz Temel, 2014, p. 766).

According to the Russian Minister of Foreign Affairs Sergei Lavrov, energy is crucial for a 'new Strategic Partnership Framework Agreement, which we regard as a vehicle for deepening our partner relations' (Lavrov, 2009) to update and replace the legal framework for EU–Russia cooperation, the EU–Russia Partnership and Cooperation Agreement (PCA) signed in 1994. Implementing a successor strategic partnership agreement is a stated priority for both the EU and Russia, yet negotiations have been delayed. Delays are in part a result of past vetoes by Lithuania and Poland and also because of the difficulties in balancing the demands of Russia as an energy producer and the EU as an energy consumer. Similar difficulties also prevented the ratification of the Energy Charter Treaty, with particular concerns on the liberalization of access to transit routes (Khrushcheva, 2011).

3. Challenges and opportunities towards the future

EU–Russia objectives for 2020 include intensifying scientific and technological cooperation, and shifting to larger commercial RES projects, yet significant obstacles remain. The first obstacle in facilitating large-scale RES trade is financial. Electricity systems need to be coupled, and significant investment will be required for grid modernization and increasing interconnection capacity (European Commission & Russia, 2013, pp. 22–23).

The second obstacle is that, as Sharples notes, there are concerns within Russia regarding how the EU's decarbonization objectives may impact Russian hydrocarbon exports to the EU (2013; Boute, 2013). In 2050, gas and oil imports into the EU could be reduced by 50 per cent or more compared with 2009 figures (European Commission, 2011a) and uncertainties regarding future gas demand are exacerbated by 'a clash of values' in EU–Russia relations in regulating the energy industry (Boussena & Locatelli, 2013, p. 188). A further obstacle is the explicit objective of Poland and Lithuania, for example, to reduce as far as possible dependence on Russian energy imports and to diversify gas supplies.

Also, political unrest in Ukraine, the annexation of Crimea and violence in the eastern part of Ukraine resulted in the rapid deterioration of EU–Russia relations between 2013 and 2014 (EEAS, 2014a, pp. 8–9). In the context of the ongoing tensions it is unlikely that Russia and the EU will take advantage of the opportunities offered by closer decarbonization cooperation. These events demonstrated Ukrainian commitment to greater European integration, which puts additional pressure on the EU to support Ukraine on its road to closer political association and economic integration (EEAS, 2014a, p. 1). This integration included the signing of the political and economic chapters of the EU–Ukraine Association Agreement in March and June 2014 respectively (EEAS, 2014b). In view of the escalating situation in Ukraine, the EU introduced three rounds of sanctions against Russia including asset freezes, visa and travel bans.

Sanctions applied to the energy sector have restricted access to the EU capital market for three major energy companies: Rosneft, Transneft and Gazprom Neft. Further measures imposed limitations on exports of certain types of energy-related technology, specifically for products for deep-water oil exploration, arctic oil exploration, and shale oil projects in Russia (Council Decision 2014/659/CFSP). The gas sector was largely unaffected by the restrictive measures in 2014, though the strained political situation raised concerns over the security of supply and demand between the EU and Russia; Russia signed a 30-year gas supply contract with China in May 2014 (Gazprom, 2014).

The 'clash of values' in traditional energy trade is not as evident in decarbonization cooperation. It is important to differentiate the impact of sanctions on EU–Russia trade in fossil fuels from dialogue on decarbonization policies. The latter is not affected by the restrictive measures directly, but the ongoing tensions could jeopardize closer decarbonization cooperation. The *potential* for cooperation on RES and energy efficiency could be considered an opportunity in the context of the uncertain future for conventional energy trade (Gazprom interview, 2010; Pichkov, 2010, Simonov, 2010), and of political relations more generally. Since both the EU and Russia are interested in continuous decarbonization (albeit to greatly different degrees), cooperation could be perceived as mutually beneficial. Trust and confidence-building measures on less politicized technical issues such as market harmonization could be a significant achievement in committing to mutual energy interdependence in an era of EU decarbonization. The feasibility of this achievement may depend on the broader context of the EU–Russia political relationship.

Russian gas will still be in high demand: with supply to China, and demand in Russia, set to increase by 0.8 per cent per year (European Commission & Russia, 2013, pp. 11–12), and as backup to intermittent EU RES in the short term at least, with European Commission 2050 scenarios predicting some continued EU gas imports (European Commission, 2011a, pp. 11–12). The EU is expected to remain the largest Russian gas (and oil) export market until about 2030 (Russia, 2010, pp. 22–23). As a result, both parties have identified carbon capture and storage technology as 'an item for further EU–Russia cooperation' (European Commission & Russia, 2013, p. 7), which could be essential for reaching decarbonization objectives. Given the lack of progress in implementing this technology (Keating, 2013; Sharmina et al., 2013), progress could be well served by EU–Russia cooperation.

There is an incomplete legal basis for decarbonization objectives in Russia, delays in the approval of regulatory acts and a lack of capital (Gusev, 2013, pp. 1–2). A legal framework for domestic Russian decarbonization objectives is needed, also for any harmonization of regulations and an increase in investment and trade between the EU and Russia. This could provide a more stable and attractive investment climate. One option would be the development of an emissions trading scheme in Russia. Emission trading systems exist in the EU, Switzerland, New Zealand and Kazakhstan, with South Korea implementing one in 2015 and pilot schemes in China in 2013 (Chen & Reklev, 2013). Given the EU's scheme, and the difficulties experienced in reaching a price on the market that encourages investment in RES and efficiency measures, the EU has the opportunity to offer expertise on best practice for such a scheme in Russia.

Another regulatory harmonization option regards targets and measures through the export of the EU's acquis on energy efficiency. However, non-binding energy efficiency targets within the EU may undermine the ability to export this part of the acquis and fail to 'provide a solution to the risks of regulatory changes that might affect energy efficiency investments in Russia' (Boute, 2013, p. 1044) or address the 'declarative nature of the Russian legislation on energy efficiency' (Boute, 2013, p. 1040). Nevertheless, the flexibility for member states and non-binding nature of the efficiency targets to 2020 and 2030 may result in lessons and best practices from individual member states.

This may provide additional input and guidance for the strengthening of the legal framework in Russia. The promotion of energy efficiency is an objective of the Roadmap on EU–Russia Energy Cooperation to 2050 and an important aspect of cooperation within the EU–Russia Energy Dialogue framework (European Commission & Russia, 2006; 2013).

Conclusions

Energy is one of key areas of EU–Russia relations. Despite the high levels of interdependence, EU–Russia energy relations are complicated due to several factors:

- A reluctance to liberalize the Russian energy sector,
- A series of Russian gas supply interruptions,
- Barriers to financing,
- And problems in the institutionalization of EU–Russia energy relations.

New political challenges faced by Europe following the political turmoil in Ukraine in 2014 are likely to contribute to further or at least continued securitization of trade in fossil fuels. We argue that despite the complicated political context of EU–Russia relations in 2014, significant potential for mutual benefits of cooperation on decarbonization exist. EU member states have expertise and experience in the development and integration of energy efficient technologies and RES, and the Russian market provides opportunities for technology transfer and trade. The EU's decarbonization objectives could stimulate the development of a domestic Russian RES industry to meet EU (and to a lesser extent Russian) demand for technology and renewable energy resources (biomass and RES generated electricity). Russia may also benefit from a 'greening' of its image, with positive effects on its standing in international environmental negotiations (Boute & Williams, 2012), increased energy efficiency and FDI including new sources of investment to modernize Russian energy infrastructure.

Cooperation on RES and energy efficiency can become an avenue to strengthening cooperation in general. Energy efficiency is a key area, described as a priority by Russia in a number of legislative acts, and is already a central aspect of the EU–Russia energy dialogue (European Commission & Russia, 2006). The question remains whether or not the EU and Russia will choose to act on these opportunities.

Despite the potential benefits of decarbonization, we also highlight that there are significant challenges to cooperation, particularly:

- A lack of legislation and problems with its implementation,
- Underinvestment,
- A lack of governmental, industry and public support for decarbonization objectives in Russia,
- The broader political context.

The EU and Russia should move from the rhetoric of cooperation on decarbonization to its implementation. This chapter suggests some specific actions from both sides, which can help to overcome the main challenges.

The EU institutions and individual member states can provide invaluable technical assistance to Russia in the development and implementation of legislation on decarbonization. The EU may consider leveraging investment in RES, energy efficiency and electricity interconnection joint projects. Commitment to the EU's 2050 decarbonization objectives will also need to be consistent, to provide both leadership and best practice, and a market for green energy technology and energy transfers.

Russia will need to demonstrate a new willingness to commit to decarbonization targets that have been set, and to implement the necessary changes to Russian legislation. These steps may include the removal of subsidies to domestic energy consumers, the liberalization of the energy sector and legislative reform. Meeting decarbonization objectives in Russia will require government leadership, alongside active industry involvement and public support. This is likely to result more from rising conventional energy costs (price security) and securing demand for cleaner energy from export partners than from a willingness to tackle the environmental costs of climate change, though this may play a salient role in the longer term.

In the short term, cooperation on energy efficiency and RES is unlikely to resolve political tension around traditional trade in fossil fuels, but may help to do so, and is predicted to be essential if a significant energy trade is to continue in a decarbonized future (by 2050). With long-term contracts to supply Russia gas to the EU, it is unlikely that EU–Russia conventional energy trade will decrease substantially before 2030. In the period to 2050, decarbonization offers a promising avenue for developing energy and broader political relations. Putting decarbonization at the centre of EU–Russia relations may be essential if a mutual interdependency is to be maintained in an era when energy systems are decarbonized, and the basis for the current hydrocarbon energy trade diminishes.

These concerns overlap with the EU's decarbonization agenda. In May 2014, the Commission stated that: 'in the long-term, the Union's energy security is inseparable from and significantly fostered by its need to move to a competitive, low-carbon economy which reduces the use of imported fossil fuels' (European Commission, 2014a). The level of success in developing a decarbonization partnership depends on the political will of the EU and Russia, as well as the broader political and economic context of EU–Russia relations. While there are significant obstacles, not least the more general political context of EU–Russia relations in 2014, the potential is there if political will supports it.

Notes

1. The EU's Energy Roadmap 2050 predicts energy import dependency could decrease from 54 per cent in 2011 to 35–45 per cent by 2050 (European Commission, 2013).
2. The Energy Charter Treaty (1994) has been signed by 51 states. The purpose of the Treaty is to provide a legal framework for international cooperation of energy industries (see www.encharter.org for list of members and observers, date accessed 20 June 2014).
3. Biofuel is a fuel (liquid, gas or solid) produced from biomass (such as agricultural waste).
4. Between October 2013 and February 2014, about 60,000 cubic metres of natural gas were consumed in Russia, 45,000 tonnes of oil and around 30,000 tonnes of coal (Russia, 2014).

Interviews

Cameron, F. (2011) EU-Russia Centre, interviewed in Brussels, Belgium, 16 February 2011.

Gazprom (2010) Anonymous representative interviewed in Moscow, Russia, 7 April 2010.

Pichkov, O. (2010) Moscow State Institute of International Relations, interviewed in Moscow, Russia, 7 April 2010.

Simonov, K. (2010) Director General of National Energy Security Fund, interviewed in Moscow, Russia, 8 April 2010.

References

Aalto, P. (2008) 'The EU-Russia Energy Dialogue and the Future of European Integration: From Economic to Politico-Normative Narratives', in P. Aalto (ed.) *The*

EU-Russian Energy Dialogue: Europe's Future Energy Security (Farnham: Ashgate), pp. 23–42.

Aalto, P. & Korkmaz Temel, D. (2014) 'European Energy Security: Natural Gas and the Integration Process', *Journal of Common Market Studies, 53,* 758–774.

Bazarova, D. (2013) 'The Wind Blows Away: Lack of Investment into Alternative Energy in Russia', *Rossiiskaya Gazeta, N6107*(131).

Belton, C. (2008) 'TNK-BP Chief is "Forced Out" of Russia', *Financial Times,* 24 July 2008.

Boussena, S. & Locatelli, C. (2013) 'Energy Institutional and Organisational Changes in EU and Russia: Revisiting Gas Relations', *Energy Policy, 55,* 180–189.

Boute, A. (2013) 'Energy Efficiency as a New Paradigm of the European External Energy Policy: The Case of the EU-Russian Energy Dialogue', *Europe-Asia Studies, 65*(6), 1021–1054.

Boute, A. & Willems, P. (2012) 'RUSTEC: Greening Europe's Energy Supply by Developing Russia's Renewable Energy Potential', *Energy Policy, 51,* 618–629.

Chen, K. & Reklev, S. (2013) 'Beijing Carbon Trading Starts as China Acts on Climate', *Reuters,* 28 November, http://www.reuters.com/article/2013/11/28/us-china-carbon-beijing-idUSBRE9AR07C20131128, date accessed 23 February 2014.

Council of the European Union (2000) 'EU-Russia Summit Joint Declaration', Paris, 12779/00 Presse, 30 October.

EDF (2013) 'EDF in Russia and the CIS', http://en-russia.edf.com/edf-in-russia-and-cis/edf-in-russia-and-the-cis-200528.html, date accessed 12 January 2014.

EEAS (2012) *Progress Report Agreed by the Coordinators of the EU-Russia Partnership for Modernisation for Information to the EU-Russia Summit of 21 December 2012* (Brussels: European External Action Service).

EEAS (2014a) *Fact Sheet on EU-Ukraine Relations* (Brussels: European External Action Service).

EEAS (2014b) *Information on the EU-Ukraine Association Agreement* (Brussels: European External Action Service).

Enerdata (2013) 'Energy Intensity of GDP at Constant Purchasing Power Parities', http://yearbook.enerdata.net/#energy-intensity-GDP-by-region.html, date accessed 10 January 2014.

Energy Policy Russia (2013a) 'Is It Necessary to Incorporate RES into the Russian Energy Strategy?', www.eprussia.ru/vote/vote.cgi?stratec=10&vote=archive&stratec=110, date accessed 12 January 2014.

Energy Policy Russia (2013b) 'Would an Increase of Energy Tariffs Encourage the RES Production in Russia?', www.eprussia.ru/vote/vote.cgi?stratec=10&vote=archive&stratec=110, date accessed 12 January 2014.

European Commission (2009) 'Delegation of the European Commission to Russia: Renewable Energy Policy and Rehabilitation of Small Scale Hydropower Plants', http://ec.europa.eu/energy/international/russia/doc/report-2009-scale-hydropower-plants-en.pdf, date accessed 11 January 2014.

European Commission (2011a) 'Energy Roadmap 2050', COM(2011) 885.

European Commission (2011b) *Energy 2020: A Strategy for Competitive, Sustainable and Secure Energy* (Luxembourg: Publications Office of the European Union).

European Commission (2012a) 'Renewable Energy: A Major Player in the European Energy Market', COM(2012) 271.

European Commission (2012b) 'EU Welcomes Russia's WTO Accession after 18 Years of Negotiations', 22 August 2012 http://trade.ec.europa.eu/doclib/press/index.cfm?id=827, date accessed 12 January 2014.

European Commission (2013) 'Energy Infrastructure: PCI Projects', http://ec.europa.eu/energy/infrastructure/pci/doc/2013_pci_projects_country.pdf, date accessed 16 December 2013.

European Commission (2014a) '2030 Climate and Energy Goals for a Competitive, Secure and Low-Carbon EU Economy', http://europa.eu/rapid/press-release_IP-14-54_en.htm, date accessed 22 January 2014.

European Commission (2014b) 'In-Depth Study of European Energy Security', SWD(2014) 330.

European Commission & Russia (2006) 'Final Report of the Thematic Group on Energy Efficiency of the EU-Russia Energy Dialogue', http://ec.europa.eu/energy/russia/reference_texts/doc/2006_10_energy_efficiency_en.pdf, date accessed 14 March 2014.

European Commission & Russia (2011) 'Common Understanding on the Preparation of the Roadmap of the EU-Russia Energy Cooperation until 2050', http://ec.europa.eu/energy/international/russia/doc/20110224_understanding_roadmap_2050.pdf, date accessed 14 January 2014.

European Commission & Russia (2013) 'Roadmap EU-Russia Energy Cooperation until 2050', http://ec.europa.eu/energy/international/russia/doc/2013_03_eu_russia_roadmap_2050_signed.pdf, date accessed 12 January 2014.

Eurostat (2013) 'EU-Russia Summit: Record Levels for Trade in Goods between EU27 and Russia in 2012', http://europa.eu/rapid/press-release_STAT-13-83_en.pdf, date accessed 3 June 2013.

Eurostat (2014) 'EU-Russia Summit: EU28 Trade in Goods Deficit with Russia Fell Slightly to 66bn Euro in the First Nine Months of 2013', http//europa.eu/rapid/press-release_STAT-14-13_en.pdf, date accessed 30 November 2014.

EU-Russia Partnership for Modernisation (2012) 'Progress Report', http://formodernisation.com/en/info/progress_report_2012.php, date accessed 9 January 2014.

Fankhauser, S. & Lavric, L. (2003) 'The Investment Climate for Climate Investment: Joint Implementation in Transition Countries', *European Bank for Reconstruction and Development, Working Paper* (77).

Gazprom (2012) 'Annual Report', http://www.gazprom.com/f/posts/55/477129/annual-report-2012-eng.pdf, date accessed 9 January 2014.

Gazprom (2014) 'Alexey Miller: Russia and China Signed the Biggest Contract in the Entire History of Gazprom', 21 May, http://www.gazprom.com/press/news/2014/may/article191451/, date accessed 20 June 2014.

Gusev, A. (2013) 'Energy Efficiency Policy in Russia: Scope for EU-Russia Cooperation', *SWP Comments*, 16(June 2013).

Hadfield, A. (2008) 'EU-Russia Energy Relations: Aggregation and Aggravation', *Journal of Contemporary European Studies*, 16(2), 231–248.

Handke S. & de Jong, J. J. (2007) *Energy as a Bond: Relations with Russia in the European and Dutch Context* (The Hague: Clingendael International Energy Programme).

IEA (2011) *World Energy Outlook 2011* (Paris: International Energy Agency/OECD).

IFC (2008) *Energy Efficiency in Russia: Untapped Reserves* (Washington, D.C.: International Finance Corporation).

IFC (2011) *Renewable Energy Policy in Russia: Waking the Green Giant* (Washington, D.C.: International Finance Corporation).

IFC (2013) *Russia's New Capacity-Based Renewable Energy Support Scheme: An Analysis of Decree No. 449* (Washington, D.C.: International Finance Corporation).

Keating, D. (2013) 'Only One Submission for CCS Funding', *European Voice*, 9 July.

Khrushcheva, O. (2011) 'The Creation of an Energy Security Society as the Way to Decrease Securitization Levels between the EU and Russia in Energy Trade', *Journal of Contemporary European Research, 7*(2), 216–230.

Kirchner, E. & Berk, C. (2010) 'European Energy Security Co-operation: Between Amity and Enmity', *Journal of Common Market Studies, 48*, 829–880.

Kulikova, E. (2010) 'Environmental and Social Standards, and the Production of Solid Biofuels in Russia', *Business I Ustoichivoe Lesopolzovanie, 2*(24), 30–36.

Laroui, F., Tellegen, E. & Tourilov, K. (2004) 'Joint Implementation in Energy Between the EU and Russia: Outlook and Potential', *Energy Policy, 32*, 899–914.

Lavrov, S. (2009) *Transcript of Remarks and Response to Media Questions by Russian Minister of Foreign Affairs Sergey Lavrov* (Moscow: Russian Ministry of Foreign Affairs), http://www.mid.ru/brp_4.nsf/0/D143CD2F32C550EBC32576770033F97C, date accessed 30 November 2014.

LUKoil (2012) 'Annual Report 2012', http, //www.lukoil.com/materials/doc/Annual_Report_2012/Lukoil_GO_2012_eng.pdf, date accessed 26 September 2013.

LUKoil (2013) 'Lukoil Promotes Wind Power Engineering', 23 January, http, //www.lukoil.com/press.asp?div_id=1&id=3761, date accessed 26 September 2013.

Maican, O.-H. (2009) 'Some Legal Aspects of Energy Security in the Relations Between EU and Russia', *Romanian Journal of European Affairs, 9*(4), 29–47.

Makarychev, A. & Meiser, S. (2014) 'The Modernisation Debate and Russian-German Normative Cleavages', *European Politics and Society*, DOI: 10.1080/15705854.2014.965897.

Maltby, T. (2013) 'European Union Energy Policy Integration: A Case of European Commission Policy Entrepreneurship and Increasing Supranationalism', *Energy Policy, 55*, 435–444.

Overland, T. & Kjaernet, H. (2009) *Russian Renewable Energy: The Potential for International Cooperation* (Farnham: Ashgate Publishing Ltd).

Pleines, H. (2009) 'Developing Russia's Oil and Gas Industry: What Role for the State?', in J. Perovic, R. Orttung & Wenger, A. (eds.), *Russian Energy Power and Foreign Relations* (New York: Routledge), pp. 71–86.

RuDanEnergo (2013) 'The Russian-Danish Energy Efficiency Center (RuDanEnergo)', http://di.dk/English/rudanenergo/about/Pages/About.aspx, date accessed 12 January 2014.

Russia (2008) *The Concept for the Long-Term Social and Economic Development of the Russian Federation to 2020, Approved by Order of the Government of the Russian Federation No. 1662-r* (Moscow: Russian Government).

Russia (2009) *Federal Law No. 261-FZ, On Energy Saving and Improvement of Energy Efficiency* (Moscow: Russian Government).

Russia (2010) *Resolution No. 1715-r on Russia's Energy Strategy until 2030* (Moscow: Russian Government).

Russia (2013a) *Mechanism for the Promotion of Renewable Energy on the Wholesale Electricity and Capacity Market, Decree 449* (Moscow: Russian Government), http://government.ru/media/files/41d469c366920ef19ca2.pdf, date accessed 29 June 2014.

Russia (2013b) *Amendments to the State Policy on Energy Efficiency, Electricity Consumption, and Renewable Energy up to 2020, Resolution N861-r* (Moscow: Russian Government), http://www.garant.ru/products/ipo/prime/doc/70288052/, date accessed 30 June 2014.

Russia (2013c) 'On the Reduction of Greenhouse Gas Emissions, Presidential Decree N 752, 30 September 2013', http://www.rg.ru/2013/10/04/eco-dok.html, date accessed 4 December 2014.

Russia (2014) 'Statistics on Energy Consumption, Ministry of Energy of the Russian Federation', http://www.minenergo.gov.ru/activity/statistic/index.php?syear=2013, date accessed 27 June 2014.

Sharmina, M., Anderson, K. L. & Bows-Larkin, A. (2013) 'Climate Change Regional Review: Russia', *Wiley Interdisciplinary Reviews, Climate Change, 4* (September/October 2013), 373–396.

Sharples, J. (2013) 'Russian Approaches to Energy Security and Climate Change: Russian Gas Exports to the EU', *Environmental Politics, 22*(4), 683–700.

Stern, J. (2009) 'Future Gas Production in Russia: Is the Concern about Lack of Investment Justified?', *Oxford Institute for Energy Studies, NG35*.

Stulberg, A. (2011) 'Moving Beyond the Great Game: The Geoeconomics of Russia's Influence in the Caspian Energy Bonanza', *Geopolitics, 10*(1), 1–25.

Tynkkynen, N. & Aalto, P. (2012) 'Environmental Sustainability of Russia's Energy Policies', in P. Aalto (ed.), *Russia's Energy Policies: National, Interregional and Global Levels* (Cheltenham: Edward Elgar Publishing), pp. 92–116.

UNEP (2014) 'JI projects', United Nations Environment Programme, 1 June 2014, http://cdmpipeline.org/ji-projects.htm#1, date accessed 12 June 2014.

US Department of Agriculture (2014) 'Russian Federation, Biofuels Annual, Biofuels Sector Update', http://gain.fas.usda.gov/Recent%20GAIN%20Publications/Biofuels%20Annual_Moscow_Russian%20Federation_6-30-2014.pdf, date accessed 30 November 2014.

Youngs, R. (2009) *Energy Security: Europe's New Foreign Policy Challenge* (London: Routledge).

Youngs, R. (2013) 'The EU's Global Climate and Energy Policies: Gathering or Losing Momentum?' in A. Goldthau (ed.), *The Handbook of Global Energy Policy* (Hoboken, NJ: Wiley-Blackwell), pp. 421–434.

11
EU–Norway Energy Relations towards 2050: From Fossil Fuels to Low-Carbon Opportunities?

Torbjørg Jevnaker, Leiv Lunde and Jon Birger Skjærseth

Introduction

The European Union (EU) has promoted ambitious energy and climate policies, particularly since 2008, aiming for progressive decarbonization towards 2050. Norway is the EU's second most important energy partner, behind mighty (and occasionally controversial) Russia. More than 30 per cent of the gas imported by Germany, France and the United Kingdom comes from Norway, and new power cables will strengthen the already important electricity trade between Norway and EU member states. Norway shares the overall climate ambitions of the EU and policymakers on both sides have cooperated closely to structure their energy interfaces so as to facilitate decarbonization.

But the EU's decarbonization ambitions also bring to light important structural challenges for this important energy interface. Fossil fuels continue to dominate the Norwegian economy. Oil and gas revenues are responsible for one quarter of the country's GNP and half its export revenues, and make up the world's largest sovereign wealth fund (currently 700 billion EUR, likely to grow to several trillion EUR by 2030). Oil and gas production causes more than a quarter of Norway's greenhouse gas (GHG) emissions. Furthermore, the share of the oil sector is increasing, leading to serious questions as to whether the country can comply with established GHG emission reduction targets. Any Norwegian government in the period to 2050 will face dilemmas of managing the key national interest of maximizing profits from the oil and gas sector on the one hand, while complying with expectations to decarbonize the economy on the other.

The EU's decarbonization target – reducing GHG emissions by 80–95 per cent by 2050 – challenges Norway's energy interests in various ways. *First, there is a market effect.* As the EU takes almost all Norwegian gas exports today, any significant reduction in European gas use will hurt the Norwegian economy. Oil is traded globally and can in theory be sold beyond Europe, but large reductions in European oil use will also represent a challenge to any Norwegian finance minister. Norwegian oil and gas interests would be challenged even further if the EU should evolve into a firm and credible global driving force for decarbonization, contributing to reduced petroleum demand globally.

Second, there is a regulatory effect. Although Norway is not an EU member and enjoys considerable policy independence, it is deeply integrated in the EU energy and environmental regulatory system through the Agreement on the European Economic Area (EEA). This agreement gives Norway access to the EU's internal market in exchange for harmonization of EU legislation relevant to such cooperation. But would any Norwegian government accept EU regulations that could negatively impact Norway's ability to extract more oil and gas from its continental shelf, and not least from the emerging Arctic petroleum bonanza?

This conventional and somewhat negatively laden national interest perspective should be complemented with a focus on how technological and political change might help Norway adapt and innovate to cope with dynamic European and global climate policies. Technological breakthrough in CCS, for instance, could greatly improve political acceptance of natural gas in the longer term. Moreover, improvements in Liquefied Natural Gas (LNG) technology might help Norway direct gas increasingly beyond Europe to balance any reduction of demand from EU member states, whether for climate policy or other reasons. Grid extension to Europe could accelerate Norway's capacity to transform its hydropower into a 'green battery' for Europe (although the potential should not be exaggerated). European clean energy and other green-tech companies would be happy takers of upscaled investments from the Government Pension Fund Global (the so-called oil fund) that might withdraw from oil and gas and increasingly target the 'decarbonization economy' instead.

Our main hypothesis, based on current policies, is that Norway will make determined efforts to maintain and secure its energy-policy independence, including the ability to maximize oil and gas proceeds from its continental shelf, even in the face of *quite* successful EU decarbonization policies towards 2050. In our main scenario, full EU decarbonization is considered highly uncertain and probably unrealistic unless the rest of the world moves in the same direction. In the following, we discuss *the prospects and conditions* for successful EU decarbonization, and the likely impacts for Norway and the Norway/EU energy interface.

Studying the implications of EU 2050 decarbonization is fraught with methodological challenges. First, uncertainties about the nature and pace of decarbonization over the coming decades loom large, and our analysis rests on assumptions that must be made as explicit as possible. Second, it is extremely difficult to distinguish impacts on Norwegian energy futures as regards

(1) EU decarbonization policies,
(2) market developments in Europe that may or may not be strongly related to EU and/or national state decarbonization policies,
(3) energy and climate policies outside the EU,
(4) market developments related or non-related to government policies beyond the EU and
(5) the possible impact of EU decarbonization policies on the rest of the world and the potential indirect impacts on Norway's scope of energy action from such EU-impacted global developments.

Tracking Norwegian impacts on the EU's decarbonization policies and practice is also challenging, although the universe of possible 'impact paths' is somewhat smaller and more manageable.

We start with an historical and contextual overview of EU–Norway energy relations, explaining how Norway is already integrated in EU energy and climate policies, and where it has retained policy independence. This section shows that Norway has been able to adapt to new EU legislation without changing practice significantly. We then present the 2014 energy and climate policy interface, and discuss what lies ahead with regard to the EU's 2020 targets. Next, we broaden the perspective. First, we present our perspective on EU decarbonization policies and scenarios to 2050, and discuss Norway's options and scope of action in relation to three main (and highly simplified) scenarios. We then turn to possible EU responses to potential Norwegian energy/climate policy scenarios, and conclude by summarizing the main arguments.

1. Historical overview

Norway has a long history of cooperating with the EU, including on energy issues. We note two interrelated types of energy relations between Norway and the EU. First, there is an economic trade relationship, where Norway is a net exporter of oil, gas and electricity, with the EU as a net importer. Despite many shared interests, this means differing interests on energy issues between Norway as a seller and the EU as a buyer, while also establishing considerable interdependence (Norges offentlige utredninger, 2012, p. 547). Second, there is a regulatory relationship through the EEA Agreement, where Norway has implemented EU energy legislation of relevance to the internal market. The EU's decarbonization strategy to 2050 is likely to affect energy relations along both dimensions, but let us first turn to how this relationship has developed historically.

1.1 Energy trade between Norway and the EU

The EU and Norway enjoy a substantial and important relationship in terms of energy trade, with Norwegian exports of oil, natural gas and electricity. Norway has supplied oil and gas to Europe since the early 1970s, gradually becoming the continent's second largest supplier (after Russia). Relatively easy to transport, oil is flexible in terms of export destination. By contrast, exports of natural gas depend mostly on pipelines built for export to countries that today are EU members. LNG is the increasingly important exception, although it constituted less than 5 per cent of Norwegian gas exports in 2013, some of which went through the Northern Sea Route to Asia (Norway, 2013b, p. 46). For the EU, imported gas from Norway covers almost 20 per cent of total gas consumption (Norway, 2013b, p. 44), and more than 30 per cent of gas imports. In 2012, Norway surpassed Russia, thus becoming the largest supplier of natural gas to the EU. The latter was partly due to a price reduction on Norwegian gas, with supply contracts gradually shifting towards spot pricing rather than the oil-pricing sustained by Russia's Gazprom (EurActiv, 2013). By early 2014, Russia was back ahead of Norway, although the 2014 political crisis over Ukraine has shrouded future export trends in considerable uncertainty (see also Chapter 10).

The EU gas market is thus extremely important for Norway, and the current trends of decreasing oil production and increasing gas production in Norway further strengthen Norway's dependence on the EU (Norges offentlige utredninger, 2012, pp. 548–549). No wonder then that Norwegian policymakers and companies are concerned about the recent squeeze on the role of gas in Europe, due to large inflows of cheap coal from the United States, massive (subsidized) growth in renewable power and energy-efficiency gains in the European power sector. These concerns are further reinforced by Russia's Ukraine adventure that provides European policymakers with another argument to turn away from gas.

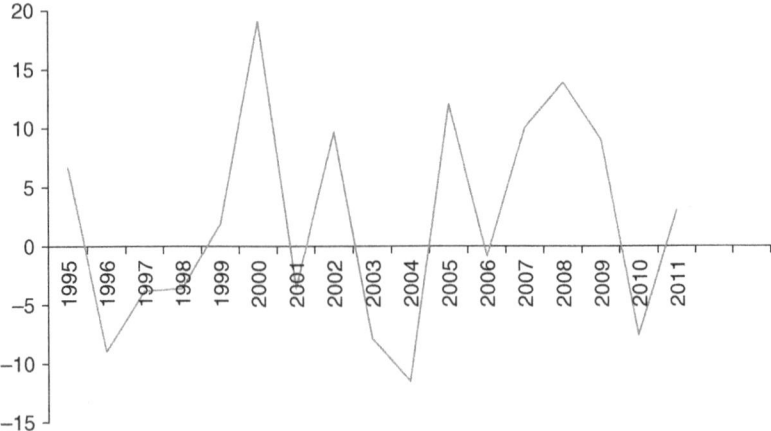

Figure 11.1   Norwegian net export of electricity 1995–2011 (in terawatt-hours, TWh)
Source: Statistics Norway (2013).

Electricity trade between Norway and the EU has also increased. Like gas, electricity trade requires a transmission network. Such cross-border network capacity has been expanded over time, first to the other Nordic countries – both prior to and after the latter joined the EU – and more recently to other EU member states as well. In 2008, an interconnector between Norway and the Netherlands was put into operation, and new interconnectors are planned to Germany by 2018 and to the United Kingdom by 2020. Norway is a net exporter of electricity. However, as shown in Figure 11.1, and in contrast to the trade in petroleum products, which are generally exported, some years have seen a net import of electricity to Norway from EU member states.

The overall significance of electricity exports is dwarfed by gas exports. In 2012, the amount of natural gas that was exported was equivalent to ten times the total Norwegian 'normal year' production of electricity (120 TWh) (Norway, 2013b, p. 44). In contrast, Norwegian power export at its highest has remained below 20 TWh (see Figure 11.1). Electricity export is likely to increase with the planned growth in renewables to 2020.

1.2   Regulation and energy relations

Energy relations between Norway and the EU extend beyond trade. An institutional framework for political cooperation that encompasses legal harmonization has been in place since 1994. Although Norway rejected EU membership in the 1972 and 1994 referenda, it participates in the EU's internal market through the EEA. Signed in 1992, this agreement remains Norway's most comprehensive international agreement. Although it grants access to the EU policy process through participation in the Commission's expert committees, the main emphasis is on Norwegian implementation of EEA-relevant EU legislation (Norges offentlige utredninger, 2012). Norway has implemented significant EU legislation through the EEA Agreement. If a legal act is deemed to be of relevance to the internal market, it is normally considered to be EEA-relevant and thus requires transposition into national legislation in the signatory countries.[1] Some adjustment of the specific legal act in question may be undertaken, but only after negotiations with the EU (Utenriksdepartementet, 2012a, pp. 16–17).

Norway has also contributed financially (3,272.5 million EUR from 1994 to 2014) to social and economic cohesion in Europe through the EEA Grants. Major parts (36.1 per cent) of these funds have been allocated to energy efficiency and renewable energy projects targeting GHG emissions (see Table 11.1).

Initially, the Norwegian government did not expect the EEA Agreement to have much impact on Norwegian energy policy (Austvik & Claes, 2011, p. 17). When the agreement entered into force in 1994, the EU had yet to develop a common energy policy, so energy issues were not given much attention. EU energy policy developed especially during the years 1994–2007. A separate 'energy dialogue' was initiated in 2002 for regular meetings between the Commission (especially its Directorate-General for Energy) and the Norwegian government (particularly the Ministry of Petroleum and Energy). This activity was envisaged as complementary to the EEA structure. The dialogue's agenda includes issues

Table 11.1 Norwegian EEA grants for energy and climate programmes 2009–2014

Theme	Amounts (percentage of total)
Environmental protection and management	€153.4 million (8.6%)
Climate change and renewable energy	€208.4 million (11.6%)
CCS	€ 184 million (10.3%)
Green industry innovation	€ 98.9 million (5.6%)

Source: Utenriksdepartementet (2012b).

concerning energy infrastructure, energy developments to 2050, natural gas, renewables and the internal energy market. The Norwegian government considers this dialogue important for mutual understanding and knowledge exchange (Utenriksdepartementet, 2012a, p. 66; Olje- og energidepartementet, 2013a, p. 2).

The geographic and competence expansion of the EU has been accompanied by a corresponding expansion of EEA cooperation. Still, Norway remains almost entirely excluded from EU decision-making, without voting rights or representation in the European Parliament, the Council of Ministers or the European Council. In energy-related policy, Norway has implemented much EU legislation of relevance to the internal market through the EEA Agreement (Utenriksdepartementet, 2012a, p. 65). EU legislation on competition and state aid, and the removal of barriers to trade has had some impacts on Norwegian energy policy (Norges offentlige utredninger, 2012, p. 547). EU legislation on licensing within the petroleum sector, on gas sales and on electricity markets has also been implemented in Norway.

However, EU legislation has been only one of several factors shaping Norwegian energy policy since the early 1990s. Other important factors include the evolution of markets, infrastructure construction and the international diffusion of general liberal economic trends (Austvik & Claes, 2011, p. 47). Moreover, Norway's position as a major energy exporter has given it energy-issue leverage vis-à-vis the EU – leverage that extends beyond limited formal access to decision-making. Thus, although EU legislation has been implemented for the petroleum sector and the mainly hydro-based electricity sector, the impact so far has been relatively weak. Successive Norwegian governments retained the scope to implement EU requirements as they see fit, which – with some exceptions – has not led to major shifts in Norwegian energy policy.

The application of EU legislation within the petroleum sector has been affected by a more general discussion of the scope of the EEA Agreement. Norway argues that the agreement does not apply to the Norwegian Continental Shelf, from which its oil and gas are extracted. Nevertheless, 'Norway has in practice yielded in the individual cases, and has given EU/EEA law effect [here]' (Norges offentlige utredninger, 2012, p. 557, our translation). Despite the underlying conflict, then, EU policies have affected Norwegian arrangements for issues such as petroleum licensing and gas sales. Preferential treatment of Norwegian companies when granting licenses for operating on the Norwegian Continental Shelf was outlawed by the EU Licensing Directive. However, major parts of such practices had anyway been removed as the result of the maturing of the Norwegian petroleum sector (Norges offentlige utredninger, 2012, pp. 548, 552) and in response to WTO requirements. Greater impact can be observed on the gas sales' structure. *Gassforhandlingsutvalget* (GFU), a centralized public body for managing Norwegian gas sales, was disbanded following an infringement process initiated by the European Free Trade Area (EFTA) Surveillance Authority (Norges offentlige utredninger, 2012, p. 548).[2] This issue has been referred to as 'one of the biggest conflicts between Norway and the EU' (Norges offentlige utredninger, 2012, p. 556, our translation), with a turbulent implementation process. However, the actual consequences were perhaps less important, as there remains a high degree of centralization in Norwegian gas exports (Claes & Vik, 2012, p. 110).

The EU deliberations to regulate the offshore energy sector, to enhance security and environmental protection in the wake of the BP oil spill in the Gulf of Mexico in 2010 (the Deepwater Horizon accident), encountered strong resistance from petroleum-production interests in Norway and oil-producing EU countries like the United Kingdom. The Norwegian minister for petroleum and energy, and the Norwegian petroleum industry, contested the EU's view that the subsequently adopted Directive for offshore safety (Directive 2013/30/EU) was relevant to the EEA (Aftenbladet, 2013).

Within the power sector, EU electricity market legislation prior to 2009 did not put much pressure on Norway, because it and the other Nordic countries were forerunners in liberalizing electricity markets in the early 1990s (Claes & Vik, 2012, pp. 108–110). However, Norwegian regulation of property rights to hydropower facilities had to be changed following a ruling of the EFTA Court. The practice of differential treatment of public and private ownership had to be removed.[3] Successive governments in Oslo were reluctant to change the Norwegian practice. Following the court ruling, a legal change was adopted that complied with the ruling by lessening the possibility of private ownership over hydropower production – thus, contrary to expectations, actually *strengthening* public ownership (Norges offentlige utredninger, 2012, pp. 560–561). Similarly, EU legislation required changes in Norway's policies on electricity consumption, but it proved possible to continue Norwegian arrangements despite complying with EU legislation.[4]

Prior to 2009, then, Norwegian energy policy did not radically change due to EU legislation. As a major energy exporter to the EU, Norway found ways to adapt without significantly altering its practice. The EU's agreement in 2009 to achieve ambitious GHG emission reduction in

2020 and 2050, however, could lead to greater impacts on Norway's energy policy – and its energy exports.

2. EU and Norway towards 2020

EU climate and energy legislation since 2008/2009 seems set to make a deeper impact on Norway. In December 2008, agreement was reached on an integrated climate and energy package. GHG emission, renewable energy and energy-efficiency targets were adopted for 2020, later accompanied by long-term roadmaps towards 2050 (see Chapter 1). EU legislation on climate change has changed radically since Norway adopted the EEA Agreement (Utenriksdepartementet, 2012a, p. 66). Prior to the EU's climate and energy package, Norwegian climate policy became more formalized and detailed due to the impact of EU law via the EEA Agreement, but EU policy served mainly to empower the positions of Norwegian actors that influenced the national policymaking process (Boasson, 2011, p. 29). EU efforts to integrate climate and energy policy by simultaneous adoption of several types of legislation challenged Norway's EEA procedures, which had been designed for handling separate legal acts from the EU (Norges offentlige utredninger, 2012, p. 565; Stortinget, 2012). Although Norway was not bound by the EU's overarching climate policy goals, the central measures in the package – revision of the EU ETS and the Renewable Energy Directive – were judged EEA-relevant and were implemented in Norway (see Table 11.2).

Norway regarded carbon pricing as the most cost-effective instrument for realizing GHG emission reduction targets. In 1991, Norway introduced a CO_2 tax, and in 2005 it established a domestic ETS. This latter became fully integrated in the EU ETS from the second trading phase (2008–2012) onwards. Norway is thus part of the revised ETS for the

Table 11.2 Norwegian transposition of the EU climate and energy package and related acts as of October 2014

EU legal act	EEA relevance	Transposition completed
ETS Directive	Yes	Yes
RES Directive	Yes	Yes
Carbon Storage Directive	Yes	Yes
Effort-Sharing Decision	No	Not applicable
Car Emissions Regulation	Yes	Pending
Fuel Quality Directive	Yes	Pending

Source: Jevnaker (2014).

third trading period (2013–2020). As the number of Norwegian sectors exposed to the EU ETS increased, the carbon price actually decreased significantly.[5] The petroleum sector (mainly offshore oil and gas production) had been subject to CO_2 taxes since 1991, but from 1998, the tax rates on oil and gas in the North Sea dropped significantly (Christiansen, 2000). An additional reduction came with the inclusion of the sector in the EU ETS in 2009. Based on the principle that total emissions costs should remain unchanged, the level of the CO_2 taxes was reduced in line with the anticipated ETS allowance price. In reality, the total carbon cost decreased significantly due to the low prices of ETS credits (Aftenbladet, 2012a), until the CO_2 tax was adjusted up again from 2013 onwards (Finansdepartementet, 2012, p. 167). Until 2020, the carbon price will probably remain too low to spur any real innovations for the longer term – the lock-in of old polluting technologies is the more likely result (Skjærseth, 2014).

In contrast to emissions trading, the Norwegian government did not welcome EU pressure for Norway to increase its use of renewable energy sources (RES). Given the already high share of RES in its energy mix (renewable hydroelectric power accounts for 96 per cent of total inland electric power consumption), this was not perceived as contributing to further GHG emission reduction. In general, energy savings and new RES will have limited impact on GHG emissions in Norway compared to most other countries. A prevalent criticism in Norway, whose climate policy has traditionally consisted of general cross-sectoral economic measures for emission reduction, was that additional GHG emission cuts in its EU ETS sectors would allow for a corresponding increase in emissions elsewhere in Europe, given the cap on total European emissions. Moreover, additional measures beyond the ETS could reduce emissions, and thereby demand for ETS allowances, in turn lessening the incentives to invest in low-carbon technology.

As regards electricity, the low capacity in cross-border networks means that increasing production could depress electricity prices. As a result, the RES Directive was subject to prolonged negotiations with the Commission. Particularly controversial was the share of renewables, which, given Norway's starting point, was expected to be set at a level far higher than for the rest of Europe (and higher than the 50 per cent maximum limit specified in the Directive). The 2020 renewables target for Norway that emerged from the negotiations was a 67.5 per cent share of total consumption, compared to 58.2 per cent in 2005 (Utenriksdepartementet, 2011, p. 5).[6] In order to increase RES-based electricity production, Norway established a technology-neutral common green certificates market with Sweden, aiming to increase RES-based power production by 26.4 TWh by 2020. In 2012, its first year of operation, this green certificates market led to an increase of 0.4 TWh in Norway (mainly hydropower) and 2.8 TWh in Sweden (Norway & Sweden, 2013). In 2014, Norway had no plans to extend the scheme beyond 2020 (*Nationen*, 2012).

The completion of Norway's transposition of other climate- and energy-related acts (the Car Emissions Regulation and the Directive on Fuel Quality, respectively) was still pending as of March 2015 (see Table 11.2). Norway has been capturing and storing CO_2 in geological structures on the continental shelf since 1996 at the Sleipner offshore field, and has worked with industry to develop CCS on gas-fired plants (Miljøverndepartementet, 2009). In 2012, Norway established a centre for development and demonstration of various CCS technologies, although the newly elected government announced a re-think in 2013 (Olje- og energidepartementet, 2013b). A full-scale, prestigious capture project at Mongstad gas-fired power plant and refinery was cancelled. Against the background of active involvement in international CCS developments as well as active engagement in the EU's legislative process, the Norwegian government considers itself as important for the EU's CCS Directive (Utenriksdepartementet, 2012a, p. 39; see also Jevnaker, 2014).

The transport-related policies in the Fuel Quality and RES Directives could put pressure on Norway. The RES Directive's sector-specific target of a 10 per cent RES share within the transport sector has been regarded as challenging, while the Fuel Quality Directive spurred controversy because the reference year against which efforts were to be measured was set at 2010. Representatives of the petroleum industry (Statoil, Esso/Exxon Mobil) and of the government criticized this for disregarding early action, that is to say, the efforts to reduce emissions prior to 2010 (Teknisk ukeblad, 2012b). Moreover, Statoil's ownership of unconventional oil production in Canada aligns its commercial interests with the Canadian position on whether there should be a common default value for calculating life cycle GHG intensity for emissions from conventional and unconventional oil production under the Directive. This would be assumed to penalize tar sands in particular, due to their uniquely high carbon content. The Norwegian Minister of Petroleum and Energy supported this critique (Aftenbladet, 2012b).

Thus, the two central pieces of the EU's climate/energy package have had opposite effects. The RES Directive forced Norway to adopt measures to increase its production of RES in line with the EU's decarbonization vision. The revised ETS seems to have weakened Norway's climate policy, due to the low carbon price that provides scant incentives for industry to invest in low-carbon solutions for the future.

3. EU and Norway towards 2050

Decarbonizing Europe by 2050 will be challenging for Norwegian petroleum-export interests, particularly in the absence of affordable CCS technologies. Despite long-term visions and binding short-term policy instruments, the future European energy mix and policy framework are highly uncertain. Long-term implications for Norway are likely to depend on at least: developments in EU climate and energy policies, technologies and energy markets; developments in international climate policy, technologies and energy markets; and the relationship between external developments and developments within Norway. In the following, we focus mainly on policy developments. Three scenarios for how this may play out are presented, followed by an assessment of how the EU could manage its energy relations with Norway under the various scenarios.

3.1 EU Policy in an international context

In March 2011, the European Commission published a roadmap prepared by DG Climate Action for moving towards a competitive low-carbon economy by 2050 (European Commission, 2011a). This low-carbon roadmap is a stepwise strategy based on GHG emission reduction milestones towards 2050. GHG emissions should be reduced by 25 per cent in 2020, 40 per cent in 2030 and 60 per cent in 2040 below 1990 levels, in order to reach a reduction of 80 per cent by 2050. Of particular relevance for Norway is, first, that electricity will play a more significant role. The power sector 'can almost totally eliminate [its] CO_2 emissions by 2050, and offers the prospect of partially replacing fossil fuels in transport and heating' (European Commission, 2011a, p. 6). Second, more domestic energy sources would be used, in particular renewables. Imports of oil and gas 'would decline by half compared to today, reducing the negative impacts of potential oil and gas shocks significantly' (European Commission, 2011a, p. 12). In June 2011, 26 of the then 27 EU member states agreed with the presidency conclusions based on the roadmap, which referred indirectly to the target of 25 per cent reduction by 2020, with Poland (supported by energy-intensive industries) blocking consensus (Skjærseth, 2013).

In December 2011, the Commission published an Energy Roadmap 2050 prepared by DG Energy (European Commission, 2011b). This energy roadmap analyses the implications of the 2050 decarbonization target, particularly for the energy sector. Here the key message is that greater energy efficiency and more use of renewables to achieve 80–95 per cent reduction by 2050 will cost about the same as continued heavy reliance on nuclear power and fossil fuels. Of relevance for Norway is that gas is set to play a key role for the transformation of the energy system. In the short and medium term, gas can substitute for coal and oil until at least 2030 or 2035. Long-term gas-supply contracts '*may* continue to be necessary to underwrite investments in gas production and transmission infrastructures' (European Commission, 2011b, p. 12, emphasis added). Oil demand will be affected by the switch to renewable and alternative fuels, but 'oil is likely to remain in the energy mix even in 2050 and will mainly fuel parts of long distance passenger and freight transport' (European Commission, 2011b, p. 12).

The future of fossil fuels beyond 2030 will depend on CCS, which will have to be applied from 2030 in the power sector in order to reach the decarbonization target. If CCS is applied on a large scale, gas may become a low-carbon technology. Otherwise, 'the long term role of gas may be limited to a flexible back-up and balancing capacity where renewable energy supplies are variable' (European Commission, 2011b, p. 12). As of 2014, EU CCS policy and deployment is in trouble for various reasons, including low public acceptance, lack of funding and a very low carbon price. At today's low carbon prices, CCS investment simply does not make sense for operators. The picture in Norway is not much brighter, following the reduced ambitions for realizing the above-mentioned prestige CCS project domestically, and announcing support to international projects as an alternative (Olje- og energidepartementet, 2014).

In June 2012, Poland vetoed the Energy Roadmap as well, arguing that EU efforts should be matched internationally and that references to 'decarbonization' should be deleted (Skjærseth, 2013). Poland is the largest coal producer in the EU, and its import dependency is among the lowest in the EU. A Polish impact assessment study of the Energy Roadmap concluded that the benefits are few and the costs high (Polish Chamber of Commerce, 2012).

These EU roadmaps have been developed as general steering documents towards 2050. To affect the energy mix in the member states, the 2050 strategies need to be backed up by binding longer-term climate and energy policies. The Commission proposed an EU climate and energy policy framework for 2030 in January 2014, adopted by the European Council in October 2014 (European Commission, 2014). The key component is binding target for at least 40 per cent domestic reduction in GHG emissions by 2030 below the 1990 level (see Chapter 1 for more details).

The Norwegian government did not take the opportunity to participate in the open consultation process in mid-2013, but sent a letter in December 2013 to the Commission stating Norway's preference for a single target – for emission reduction – for EU decarbonization plans to 2030. The Norwegian contribution to decarbonization was presented as providing electricity for capacity markets, and through continued exports of natural gas, which emits less CO_2 than other fossil fuels (Norway, 2013a).

Adopting new climate and energy policies will also depend on international developments. Even though unilateral EU action can provide various benefits such as early-mover advantages, reduction of import dependencies and significant air pollution and health co-benefits, decarbonization scenarios assume that *global* climate action is taken. In particular, global action is needed to counter carbon leakage and adverse effects on competitiveness in energy-intensive industrial sectors exposed to international competition. Moreover, the costs of investment depend on global regulatory frameworks, and most energy technological developments are likely to emerge in Asia and the USA. This means that achieving a new, binding and ambitious climate agreement in 2015 (or later) will be pivotal for decarbonizing Europe, and consequently for the implications for Norway.

The 2008 climate and energy package was intended to strengthen EU leadership towards a new, ambitious and binding climate treaty that would include all major emitters. That failed, as nations like the United States, China and India refused to follow. The 2009 'Copenhagen Accord' did not set global reduction targets, did not amount to what scientific advice holds is necessary to remain within the 2°Celsius objective, and it was not legally binding. Since then, progress on a new global climate agreement has been slow. Even with the 2012 Doha outcome including a second commitment period under the Kyoto Protocol applicable to the EU and Norway, it is highly uncertain whether a new climate agreement can align EU actors in practical support of decarbonizing Europe by 2050. In the 2030 framework, the European Council has not proposed to raise the 40 per cent target if others follow, but will revert to the issue after the Paris Conference. An increase in ambition is highly unlikely. This change in strategy reflects the poor experience from Copenhagen and increased opposition to unilateral action within the EU.

3.2   Norway's options in the EU/international context towards 2050: Three scenarios

Based on the potential developments within the EU and at the international level, we have synthesized three scenarios. In the *first scenario*, the outcome in Paris will be based on a 'pledge-and-review' approach. The EU's 2030 plan fails or become significantly watered down. Norway will then be left with the current 2020 package of policies. In that case, the outcome is fairly predictable: increase in RES production based on more hydroelectric power and some wind power, in line with the 2009 RES Directive. The certificate system and other subsidies are not likely to continue after 2020. Still, the EU's RES Directive is expected to contribute to Norwegian and Nordic surplus of electricity in 2020 (Teknisk ukeblad, 2012c; 2012d). A lively debate in Norway questions whether the surplus should be exported (if this would in fact contribute to reducing total emissions in Europe, given the overall cap set by the EU), or if it should be used to replace the use of fossil fuels in the transport and petroleum sectors (Teknisk ukeblad, 2012a). This issue is far from resolved, but the change in installed capacity might trigger a stronger push for exporting electricity from the Nordic power market to other parts of Europe, thus expanding Norway's electricity relations with the EU after 2020 and bolstering Norway's role as net exporter of clean electricity.[7]

The carbon price is likely to remain low until 2020, depending on economic activity and developments in emissions in the EU ETS sectors. This will ease the pressure on the petroleum industry and make it more difficult to develop commercially available CCS. We will generally not see the EU directly challenging Norwegian resource management in the Arctic on climate change grounds. The pressure for further exploration for Arctic oil and gas is likely to increase, although more technologically advanced and expensive projects will be sensitive to price developments in the oil and gas markets.

In the *second scenario*, the outcome from Paris will be somewhat more than business as usual. The Commission's 40 per cent GHG cut proposal for 2030 will be adopted with a 'view' to 2050 decarbonization. The implications for RES production in Norway are likely to be as above. Pressure will increase on the EU ETS sectors, including the petroleum industry. If the carbon price rises significantly, incentives to develop CCS and other low carbon technologies will increase. Norwegian gas interests will benefit from higher carbon prices, as coal has higher CO_2 emissions than gas. In 2014, coal was outperforming gas, due to the low carbon price, low coal price and high gas price. The price difference between gas and coal is also linked to the imperfect liberalization of the European gas market. Increased liberalization, including moving from long-term oil-indexed contracts towards spot pricing, will curb the gas price, which is not necessarily in line with Norwegian gas industry interests.

In this scenario, Europe is likely to continue to demand Norwegian gas and oil (while perhaps less than in 2014), and Norway will probably also have succeeded in diversifying gas exports in the form of LNG to other less decarbonization-prone regions (gas prices in East Asia are assumed to remain far higher than in Europe). At the same time, ambitious EU policies are likely to be mirrored in quite ambitious Norwegian climate policies. These policies will probably reflect current Norwegian priorities: investing in CCS, focusing on cost effectiveness, working with EU and other likeminded players for effective ETS markets and a higher carbon price, investing in regions and in areas of low-hanging fruits of carbon reduction through tropical forest protection, and standing forth as a major financial investor in decarbonization (not least through its oil fund). It is also likely in this scenario, which assumes a gradually higher carbon price, that the most risky and expensive fields, including some in Arctic areas, will not be exploited, because higher carbon prices will make them unprofitable. On the whole, oil and gas are likely to lose some of their current primacy in the Norwegian economy, although retaining an important role.

In the *final scenario*, the EU makes its decarbonization vision and GHG emission reduction trajectory towards 2050 legally binding, with massive support within an ambitious international climate treaty based on global carbon pricing. This will prove extremely challenging for Norway's petroleum interests. It will strengthen the environmental lobbies calling for a halt to any further exploration for Arctic oil and gas and to some developments in mature areas on the Norwegian continental shelf. The argument that scientific analyses indicate that new Arctic fossil fuel reserves should remain in the ground to allow the world to achieve the 2°Celsius target was actively pursued by green parties in the 2013 Norwegian election campaign, and has solid resonance in similar quarters across Europe, with calls for a moratorium on Arctic petroleum exploration.

Whether climate change concerns will impact Norwegian petroleum policies in the High North (which they have not yet done) are likely to be decided by global market developments influencing the profitability of Arctic drilling (that is, the shale-gas revolution). Global climate policies can be one factor, if nations agree on a stiffer carbon price. If the EU continues to press for more radical climate policies globally, that may come to constitute one of several forces that tip the economic balance in the Arctic against further production of oil and gas. That would also strengthen purely economic arguments in favour of reduced exploration as a result of the rise in exploration costs, declining export markets and domestic macroeconomic concerns. Recently, a government-appointed committee consisting of representatives from the Ministry of Finance, the two largest employers' and trade unions and Statistics Norway argued that Norway should slow down its exploration and development of new fields for economic reasons.[8] The government promptly dismissed this proposal. In a related development, however, the government decided in March 2014 to solicit a comprehensive study of the pros and cons of moving the 700 billion EUR oil fund out of fossil fuels (which currently comprise almost 10 per cent of the fund's investments). Should this option get support in the upcoming study, it will boost investment in non-fossil fuel sectors globally, and should serve to question the predominance of oil and gas in the Norwegian economy.

The main option for maintaining the Norwegian fossil fuel regime appears to be the development and commercialization of CCS. Norway has, as noted above, invested significant resources in CCS development applied to gas, albeit with limited success. We can expect the development of CCS to get a forceful push under this scenario, as most fossil fuel-exporting actors will find themselves facing the same challenge.

3.3   EU responses and management

The main challenges for the EU are threefold. First, Norway strongly emphasizes the importance of retaining its national sovereignty over its energy mix, strategy and policy. A second and related point is that Norway is not an EU member. It has an EU-sceptical population and few decision-makers who consider Norway in an EU context. Finally, there is a split between the Norwegian Ministry of the Environment, which has main responsibility for climate policy, and the Ministry of Petroleum and Energy. This split, at least partially mirrored in conflicts between environmental and energy/industry players in EU institutions, impedes the government's ability to develop unified Norwegian positions and makes it more difficult for the EU to manage its relations with Norway towards decarbonization.

Despite these challenges, how the EU will manage its energy relations with Norway depends on the scenario. In the first scenario, existing cooperating structures described above are not likely to change significantly. The EU is managing its evolving energy relations with Norway by keeping an open door based on buyer/seller interdependence. In the second scenario, investment in new infrastructure and the renegotiation of long-term gas supply contracts will be challenging for Norway. This may call for more active coordination of EU external energy relations. As to the final scenario, the EU 2050 Energy Roadmap states explicitly that the transition to a low-carbon economy will be managed in close contact with the EU's energy partners, including Norway. Due to the importance of the Norwegian energy sector for the EU, Norwegian decision-makers enjoy good access to the European Commission, particularly DG Energy, and other EU institutions. Moreover, closer cooperation could arise through coordinating the implementation of EU decarbonization policy, due inter alia to the need to flesh out the technical details in expert committees. Participation in policy-preparation of authorities and stakeholders from Norway could also strengthen cooperation. Nevertheless, although access to deliberations is often granted, this cannot compensate for the lack of the right to vote, as discussions are often made under the 'shadow of the vote' – arguments are weighed in light of voting power. If the EU were to move towards stronger policy measures to achieve decarbonization by 2050, this could increasingly put pressure on the 'sustainability' of Norway's association with the EU through the EEA Agreement. However, the 2014 Ukraine crisis may increase the attractiveness of Norway as a fossil fuel provider to the EU as increased delivery of oil and gas from Norway could reduce EU dependence on energy imports from Russia (Aftenposten, 2014).

Conclusions

The key EU challenge involved in engaging Norway in decarbonization lies more in the country's fossil fuel interests than in how external energy relations are managed by the EU. The transformative effect of oil and gas has been significant in the Norwegian economy. Essentially, Norway today is locked in a fossil regime where resources, knowledge, policy, technology, organizations and business are directed towards the exploration, production, distribution and consumption of fossil energy. The key challenge is to transform this fossil energy regime into a regime based on low-carbon energy and solutions.

Against this backdrop, we have explored three scenarios based on EU and international climate policies that to varying degrees may challenge the Norwegian fossil fuel regime and offer new low-carbon opportunities. In the final and 'greenest' scenario, the EU makes its decarbonization vision and GHG reduction trajectory towards 2050 legally binding, backed by massive internal support within an ambitious international climate treaty based on global carbon pricing. We believe that *this* is what it will take to seriously challenge Norway's petroleum interests in the absence of domestic reasons for changing pathways. Still, external pressures also contribute by providing new opportunities for Norway, like the export of renewable electricity, the development of low-carbon technologies and internal debates on disentangling the 700 billion EUR oil fund from fossil fuels.

Fossil fuels constitute an important dimension for Norway's relations to the EU and not least major EU member states like the United Kingdom and Germany. Issues related to trade in petroleum products constitute a major share of Norwegian interaction with Europe. This concerns interaction within the political and economic realm, which perhaps would be significantly lower without such trade. Consequently (in the absence of successful CCS), decarbonization challenges not only the regime of fossil fuels, but also a major part of EU-Norwegian relations. To avoid a weakening of these ties, the political and economic bonds that accompany trade in petroleum products will have to be replaced by something else. While increasing trade in RES-based electricity could represent an interesting avenue for future energy relations under decarbonization, even the prospect of Norway as a 'green battery' is unlikely to offer an equivalent replacement given the sheer magnitude of Norway's petroleum-related export to the EU.

Notes

1. This occurs through a screening process. The EEA Committee, consisting of representatives from the signatory countries and the EU, evaluates separate EU legal acts for 'EEA relevance'.
2. Implementation of EU legislation in Norway is monitored by the EFTA Surveillance Authority, which can bring a case of Norwegian non-compliance before the EFTA Court (Utenriksdepartementet, 2012a, p. 17).
3. Public owners were granted never-ending licenses, whereas private owners had to return ownership to the state when licenses expired *['hjemfall']*.
4. Long-term contracts granting cheap electricity to energy-intensive industries were replaced due to EU regulations. Norway subsequently established a system of guarantees provided by the state, with a market-based premium paid by the companies. Moreover, the exemption from the electricity tax in some geographical areas had to be removed as it violated EU competition law – although the practice was maintained by setting the general tax rate low enough to make the exemptions acceptable under EU state-aid rules (Norges offentlige utredninger, 2012, pp. 561–562).

5. The EU ETS encountered major problems due to the economic and financial crises and generous import of external credits, leading to an over-allocation of allowances and a very low carbon price.
6. Roughly put, the share is calculated by dividing RES production and RES consumption with overall energy consumption. Energy consumption within energy producing sectors is not included in the later figure. Thus, the use of natural gas within the petroleum sector is excluded from the calculation, which would have significantly lowered the Norwegian RES share (Bøeng, 2012, p. 8).
7. One idea is to use hydro reservoirs as batteries that can be 'charged' in periods of surplus electricity production from intermittent RES. This could be done via pumped storage, in which the electricity is used to pump water to a higher reservoir. Alternatively, output from hydropower plants with reservoirs could be reduced or saved for balancing purposes. While the technical potential is estimated to be rather high (varying between 10,000 MW and 25,000 MW in Southern Norway) (Teknisk ukeblad, 2010), this must be coordinated with the rest of the power system, so the commercial potential is significantly lower (Teknisk ukeblad, 2012e).
8. One argument is that Norway's dependence on petroleum revenues requires a controlled reduction in activities to alleviate the expected drop in future revenues (Norges offentlige utredninger, 2013).

References

Aftenbladet (2012a) 'CO$_2$-utslepp er Blitt Billigare,' 30 January, http://www. aftenbladet.no/klima/CO2-utslepp-er-blitt-billigare-2924197.html#.U1Z9_vl_uCk, date accessed 22 April 2014.

Aftenbladet (2012b) 'Utsetter Oljesand-nei til 2013', 22 April, http://www. aftenbladet.no/energi/klima/Utsetter-oljesand-nei-til-2013-2961808.html, date accessed 29 July 2013.

Aftenbladet (2013) 'Norge Avviser Offshore-Direktiv, EU Har Døren På Gløtt', 22 May, http://www.aftenbladet.no/energi/Norge-avviser-offshore-direktiv_-EU-har-doren-pa-glott-3182211.html#.U1Z0ivl_uCk, date accessed 22 April 2014.

Aftenposten (2014) 'Intense Energi-samtaler Mellom Norge og EU', 7 May, http://www.aftenposten.no/nyheter/iriks/Intense-energi-samtaler-mellom-Norge-og-EU-7557886.html, date accessed 30 October 2014.

Austvik, O. G. & Claes, D. H. (2011) 'EØS-avtalen og Norsk Energipolitikk' *Europautredningen*, Rapport 8 (Oslo: Europautredningen).

Boasson, E. L. (2011) 'Norsk Miljøpolitikk og EU: EØS-avtalen som Inspirasjonskilde og Maktmiddel' *Europautredningen*, Rapport 19 (Oslo: Europautredningen).

Bøeng, A. C. (2012) 'Fornybardirektivet – Hva betyr det for Energibransjen?' Report written *by Statistics Norway* on behalf of *Energy Norway*.

Christiansen, C. A. (2000) 'On the Effectiveness of Environmental Taxes: The Impacts of CO$_2$ Taxes on Environmental Innovation in the Norwegian Petroleum Sector', *FNI Report 10/2000* (Lysaker: Fridtjof Nansen Institute).

Claes, D. H. & Vik, A. (2012) 'Kraftsektoren: Fra Samfunnsgode til Handelsvare', in D. H. Claes & P. K. Mydske (eds.), *Forretning Eller Fordeling: Reform av Offentlige Nettverkstjenester* (Oslo: Universitetsforlaget), pp. 98–121.

EurActiv (2013) 'Norway Overtakes Russia as EU's Biggest Gas Supplier', 25 June, http://www.euractiv.com/energy/norway-overtakes-russia-biggest-news-528854, date accessed 13 August 2013.

European Commission (2011a) 'A Roadmap for Moving to a Competitive Low Carbon Economy in 2050', COM(2011) 112 final.

European Commission (2011b) 'Energy Roadmap 2050', COM(2011) 885/2.

European Commission (2014) 'A Policy Framework for Climate and Energy in the Period from 2020 to 2030', COM(2014) 15.

European Parliament (2012) 'Parliament Calls for Low-carbon Economy by 2050', PR\40876.

Finansdepartementet (2012) 'Skatter, Avgifter Og Toll 2013', Prop. 1 LS (2012–2013).

Jevnaker, T. (2014) 'Europeanization or Cherry-Picking? Norway's Implementation of the EU Climate and Energy Package', *FNI Report* (Lysaker: Fridtjof Nansen Institute).

Miljøverndepartementet (2009) 'Norwegian Climate Policy: Carbon Capture and Storage (CCS)', 2 December, http://www.regjeringen.no/, date accessed 15 August 2013.

Nationen (2012) 'Regjeringen Vil Fjerne Grønne Sertifikater i 2020', 18 October, http://www.nationen.no/, date accessed 16 August 2013.

Norges offentlige utredninger (2012) 'Utenfor og Innenfor: Norges Avtaler Med EU', *NOU, 2012*(2).

Norges offentlige utredninger (2013) 'Lønnsdannelsen og Utfordringer for Norsk økonomi', *NOU, 2013*(13).

Norway (2013a) 'Letter to the European Commission, and to the Commissioners for Climate Action, and for Energy, Ministry of Foreign Affairs, Ministry of the Environment, and Ministry of Petroleum and Energy', 19 December.

Norway (2013b) *FACTS 2013: The Norwegian Petroleum Sector* (Oslo: Ministry of Petroleum and Energy).

Norway & Sweden (2013) *Et Norsk-Svensk Elsertifikatmarked: Årsrapport for 2012* (Oslo/Stockholm: Norges vass- og energidirektorat & Energimyndigheten).

Olje- og energidepartementet (2013a) 'Non-Paper: Norwegian Views on European Energy Issues', http://www.regjeringen.no/upload/OED/Foredrag%20i%20pdf%20som%20er%20lagt%20ut%20p%C3%A5%20nettet/Norway_non-paper_20130522.pdf, date accessed 8 October 2014.

Olje- og energidepartementet (2013b) 'Ønsker ny Vurdering av CO_2-Håndtering', 1 November, http://www.regjeringen.no/nb/dep/oed/pressese nter/pressemeldinger/2013/onsker-ny-vurdering-av-co2-handtering.html?id=744870, date accessed 2 November 2013.

Olje- og energidepartementet (2014) 'Prop. 1 S (2014–2015) Proposisjon til Stortinget for Budsjettåret 2015', http://www.statsbudsjettet.no/dokumenter/gulbok.pdf, date accessed 8 October 2014.

Polish Chamber of Commerce (2012) *Assessment of the Impact of the Emission Reduction Goals Set in the EC Document 'Roadmap 2050' on the Energy System, Economic Growth, Industry and Households in Poland* (Warsaw: EnergSys).

Skjærseth, J. B. (2013) 'Unpacking the EU Climate and Energy Package: Causes, Content, Consequences', *FNI-Report 2*(2013) (Lysaker: Fridtjof Nansen Institute).

Skjærseth, J. B. (2014) 'EU Emissions Trading: Achievements, Challenges, Solutions', in T. L. Cherry, J. Hovi & D. M. McEvoy (eds.), *Towards a New Climate Agreement* (London: Routledge), pp. 254–265.

Statistics Norway (2013) 'Electricity: Annual Figures 2011', http://www.ssb.no/en/energi-og-industri/statistikker/elektrisitetaar/aar/2013-0320?fane=tabell&sort=nummer&tabell=104211, date accessed 3 January 2014.

Stortinget (2012) 'Møte i Europautvalget Torsdag den 18. Oktober 2012 kl. 08.30', http://www.stortinget.no/, date accessed 30 October 2014.

Teknisk ukeblad (2010) 'Pumpekraft: Toget Kan Gå Fra Oss', 1 September, http://www.tu.no/kraft/2010/09/01/pumpekraft–toget-kan-ga-fra-oss, date accessed 13 June 2014.

Teknisk ukeblad (2012a) 'Grønn Bløff Eller Klimatiltak?', 31 August, http://www.tu.no/meninger/tumener/2012/08/31/gronn-bloff-eller-klimatiltak, date accessed 16 August 2013.

Teknisk ukeblad (2012b) 'Norske Raffinerier Kan Straffes For Klimatiltak', 24 February, http://www.tu.no/olje-gass/2012/02/24/norske-raffinerier-kan-straffes-for-klimatiltak, date acccessed 29 July 2013.

Teknisk ukeblad (2012c) 'NVE Forsvarer Elsertifikatene', 31 August, www.tu.no/energi/2012/08/31/nve-forsvarer-elsertifikatene, date accessed 16 August 2013.

Teknisk ukeblad (2012d) 'Vi Må Ha Full Fart På Utbygging Av Kablene', 19 September, http://www.tu.no/energi/2012/09/19/-vi-ma-ha-full-fart-pa-utbygging-av-kablene, date accessed 16 August 2013.

Teknisk ukeblad (2012e) 'Statnett Knuser Pumpekraftdrømmen', 16 November, http://www.tu.no/kraft/2012/11/16/statnett-knuser-pumpekraftdrommen, date accessed 13 June 2014.

Utenriksdepartementet (2011) *Samtykke Til Deltakelse i en Beslutning i EØS-komiteen om Innlemmelse i EØS-avtalen av Direktiv 2009/28/EF om å Fremme Bruken av Energi fra Fornybare Kilder (Fornybardirektivet)*, Prop. 4 S, (Oslo: Government of Norway).

Utenriksdepartementet (2012a) *EØS-avtalen og Norges øvrige avtaler med EU: Sentrale Prioriteringer og Virkemidler i Norsk Europapolitikk*, Meld. St. 5 (2012–2013) (Oslo: Government of Norway).

Utenriksdepartementet (2012b) *EØS-midlene: Solidaritet og Samarbeid i Europa*, Meld. St. 20 (Oslo: Government of Norway).

12
Conclusions: Lessons Learned

Claire Dupont and Sebastian Oberthür

In this volume, we explored the implications of the EU's objective to reduce greenhouse gas (GHG) emissions by 80–95 per cent by 2050, compared to 1990 levels, on internal and external EU policies and strategies. The GHG emission reduction goal was agreed at the political level by the European Council of heads of state and government of EU member states in October 2009 (European Council, 2009). The societal transformation implied in such a goal is summed up in the term 'decarbonization'. Aiming for decarbonization responds to calls from the global scientific community (most notably expressed through the periodic reports of the Intergovernmental Panel on Climate Change) to reduce dramatically our GHG emissions as soon as possible to avoid potentially catastrophic consequences of human-induced climate change (IPCC, 2013). All contributors to this book approached their particular topic with the decarbonization goal as the main point of reference of their analysis.

Chapters 2–7 discussed a number of key sectors of internal energy-related policy from the perspective of achieving decarbonization. Authors focused on the questions:

(1) how far has the EU come along the pathway towards decarbonization in various policy sectors;
(2) how much work remains to be done; and
(3) what policy options are available for closing the gap, considering the drivers and barriers identified?

Within many of the policy sectors analysed, detailed scenarios for achieving decarbonization already exist and provided essential data for authors. The EU's own roadmaps include an overarching roadmap on moving towards a low-carbon economy (European Commission, 2011b), a roadmap for the energy sector (European Commission, 2011a) and a roadmap for the transport sector (European Commission, 2011c). These and other scenarios or roadmaps discussed in Chapter 1 provide important background for any policy evaluation and design. By examining functional overlaps between sectoral policies and the overarching decarbonization goal, political will (and long-term planning), institutional set-up and societal involvement, authors were able to shed light on some of the reasons for the progress made thus far and to identify toeholds for moving forward.

Chapters 8–11 turned attention to the implications of the EU's decarbonization goal on its external relations. They thus extended the analysis of this book beyond the borders of the EU to explore how decarbonization is affecting, or could or should affect, external relations with a number of energy partners, especially in the near abroad. With decarbonization as the central focal point, authors could envision a number of scenarios for how such relations may evolve, while pointing to the challenges and opportunities decarbonization raises in these cases. By including this external dimension, the book addresses an aspect of the EU's drive towards decarbonization that has so far received scant attention, despite – as the contributions to the volume amply demonstrate – its richness and importance.

In this concluding chapter, we synthesize the analysis and results of this volume by presenting nine key findings, including some guidance for future policy development, strategic planning and further research.

1. The EU has made important progress towards decarbonization

The significant progress made by the EU towards decarbonization is evident from the GHG emission reduction the EU has achieved so far. As discussed in Chapter 1, GHG emissions in the EU-28 had declined by about 19 per cent compared to 1990 levels by 2012 (see Figure 1.1). Available preliminary data for 2013 and early member state estimates for 2014 indicate that emissions have continued to decline, so that emissions by 2020 are expected to exceed the 20 per cent reduction target (EEA, 2014). The EU has thus achieved emission reductions considerably beyond its international commitments. It by far exceeded its commitment under the Kyoto Protocol to reduce its GHG emissions by 8 per cent on average between 2008 and 2012, compared to 1990 levels. It is also set to overachieve, by an as yet uncertain (but possibly considerable) margin, its commitment to a GHG emission reduction of 20 per cent by 2020, enshrined internationally in the second commitment period under the Kyoto Protocol from 2013 to 2020 (EEA, 2014).

The EU may thus be seen to be leading the way internationally on reducing GHG emissions and towards decarbonization by 2050. Not only has the EU repeatedly come out with international commitments to reduce its emissions in future, but it has also already overachieved its past international commitments. This may reinforce the EU's credibility in pledging to reduce its GHG emissions by 40 per cent by 2030 (also compared to 1990 levels) as input to the international negotiations that are expected to lead to a new international agreement in Paris in 2015. The progress made by the EU in reducing its emissions is all the more important given the slow international negotiations and few comparable commitments to reduce GHG emissions in partner countries (Eckersley, 2012). By taking on the highest reduction target among major emitters, the EU's pledge for 2030 sends a positive signal for further progress towards decarbonization (Oberthür & Wyns, 2014), even though, as of early 2015, the EU has yet to elaborate policy measures to ensure the 2030 goal will be achieved.

The decline of GHG emissions in the EU has been underpinned by the development of policy measures that arguably amount to the most comprehensive climate policy framework of major countries worldwide. The EU has been developing climate policy internally since the 1990s, not least driven by its ambitions to lead the world on stringent measures to reduce GHG emissions. Targeted policy measures in the energy sector, especially, include the EU Emissions Trading System (ETS) covering emissions from large installations, energy efficiency measures for products, vehicles, services and buildings and policies to promote the development of renewable energy (see Table 1.1 and overview in Chapter 1). These policies have, to a significant extent, driven the emission reductions achieved (see, for example, EEA, 2014), even if other factors (such as the financial and economic crises from 2008 onwards) also contributed. The role and importance of targeted policy measures towards decarbonization is further reinforced by the analyses in Chapters 2–7 that illustrate critical differences among subsectors (as further discussed below).

2. With EU 'catch-up governance', progress remains insufficient

Despite the progress the EU has made in reducing GHG emissions, the analysis in this book indicates that such progress has remained insufficient to achieve decarbonization by 2050. Running behind schedule, progress can be characterized as 'catch-up' governance. The successive steps of policy strengthening have consistently lagged behind what is required. The 2030 climate and energy framework, agreed by the European Council in October 2014, continued that pattern and has put off much of the effort required to later decades. EU climate and energy policy has thus produced insufficient guidance and incentives for the cross-sectoral transformation called for by the decarbonization goal. If the pattern of 'catch-up governance' continues, the EU may not achieve its long-term decarbonization objectives.

While important challenges on the road to decarbonization remain in all policy areas investigated in this volume, progress in emission reductions and policy development has also varied. The power sector has perhaps seen the greatest progress among the sectors in focus, even though the remaining challenges are enormous. GHG emissions in the sector declined by 22 per cent between 1990 and 2013 (EEA, 2014, p. 45), and the potential for this sector to eliminate most GHG emissions is well acknowledged (European Commission, 2011b). Important strides have been made in establishing a firm European policy framework for the sector, including the EU ETS, the Renewable Energy Directive as well as the *acquis communautaire* to advance energy efficiency (Energy Efficiency Directive, etc. – see Table 1.1). However, this undeniable progress has so far remained insufficient from a 2050 perspective. To achieve decarbonization, the European Commission envisions that the power sector must reduce emissions by between 93 and 99 per cent by 2050 (European Commission, 2011a; see Chapter 3). At the same time, demand for electricity is expected to grow as electricity replaces fossil fuels in other sectors, including industry and transport (see Chapters 5 and 6). Decarbonizing the power sector requires not only a shift from fossil fuel electricity generation, but also structural changes in the electricity grid to increase interconnections and cope with variable sources of energy (such as wind and solar). However, the internal electricity market and the upgrade of the electricity grid are facing delays, while the further roll out of renewable energy sources faces uncertainty, both regarding the appropriate grid infrastructure and the stability of the regulatory framework (Chapters 2–4).

Progress in the other sectors investigated has been less pronounced and/or is facing even greater challenges. In the industrial sector investigated in Chapter 5, GHG emissions have been declining since 1990, with the majority of these reductions due to improved efficiency measures, but further reduction potential seems uncertain at best. The energy intensive industry faces particular challenges under decarbonization. Moving towards more (low-carbon) electricity-based processes provides only a partial solution. Industry is expected to reduce emissions in 2050 by at least 83–87 per cent, but the technological solutions for energy intensive industries to achieve this goal remain underexplored and underdeveloped. An overarching climate policy framework for the sector was established with the EU ETS. However, low-emission allowance prices have provided a poor incentive for stimulating research and development into required breakthrough technologies. Allocating free emission allowances under the ETS (for reasons of international competition) has only weakened any such incentive further. There is a need for the EU to provide a more enabling framework for research and development sooner rather than later, as long investment cycles in heavy industry (20–40 years) mean that solutions need to be put in place decades before 2050 to achieve the goal. In this particular sector, policy lags far behind the required timetable for action (Chapter 5). Agreements reached by the European Council in October 2014 to set aside some funds from 2020 to support related innovation may provide some perspective for improvement.

In the transport sector, GHG emissions have even been moving in the wrong direction. Improvements in vehicle efficiency have been outpaced by the growth in the number of vehicles (road, maritime and air, in particular) travelling further distances, leading to emissions from the transport sector in the EU increasing by 29 per cent between 1990 and 2009, although emissions began to decline slightly after 2007 (see Chapter 6; Hill et al., 2012). By 2012, the transport sector represented a quarter of total EU GHG emissions (European Commission, 2014, p. 127), and emission growth is expected to return if no further policy measures are put in place. As the discussion in Chapter 6 illustrates, policies addressing the problem have been evolving, with a main focus on vehicle efficiency and emission standards. However, these vehicle standards lack stringency to address effectively the problem and will likely need to be complemented with other less established policy strategies and instruments (including modal shift, pricing schemes, integrated city planning and so on). The electrification of transport is one viable option for reducing direct emissions of GHGs, but this requires the aforementioned shift in the power sector and electricity infrastructure. As with the power and industry sectors, the transport sector requires further strategic planning and investment in infrastructure and technological development to achieve decarbonization by 2050 (see Chapter 6).

Progress with respect to buildings has been similarly precarious. GHG emissions are linked to the high energy consumption of buildings, which has remained at about 40 per cent of total EU final energy consumption for several decades (BPIE, 2011; European Commission, 1998). Progress in improving overall energy efficiency in the EU has occurred, with final energy consumption of the EU being 7.3 per cent lower in 2012 than in 2005 (EEA, 2014, p. 76), but the building sector's contribution to this progress has been marked by untapped potential. While an encompassing policy framework for improving the energy performance of buildings was established in the EU with the adoption of the Energy Performance of Buildings Directive in 2002, the results have not been impressive, given the need to drastically reduce emissions. Member states, in particular, have consistently watered down policy proposals during the policymaking process. These weaker measures have then been poorly implemented, since member states reserved much flexibility for themselves (Chapter 7). This weakness contrasts with the huge technological potential for reducing the building stock's energy consumption to nearly zero, thus almost eliminating buildings' contribution to GHG emissions. Such technological solutions include (among others) improved efficiency of buildings materials, insulation and building techniques, and on-site renewable energy installations. Hence, it is the sectoral policy framework that has not yet allowed the building sector to play a more significant role in moving towards decarbonization.

The EU seems to be caught in a cycle of insufficiently ambitious past policy actions, problems of implementation and incremental advances to fix such policies, engaging in what we have termed 'catch-up governance'. Decisions on large-scale infrastructure to put the EU on the road to decarbonization need to be taken sooner rather than later (Neuhoff et al., 2014; see Chapters 1–7). Power plants, infrastructure projects, buildings and industrial installations have long operational lifetimes, meaning that investment decisions taken today influence the EU's ability to achieve its 2050 objective. Decarbonization calls for steadily advancing ambition backed up with strategic policy measures that aim to change the very structure of society, well in advance of 2050.

3. There is further need for long-term, strategic planning

Policymaking for decarbonization by 2050 requires a long-term perspective. In each of the chapters contributing to this volume, a number of win-win situations are highlighted in the move to decarbonization that also provide reasons for taking action on climate change in the short to medium term. For example, reducing GHG emissions in power, transport, industry and buildings results in lower air pollution, improved efficiency, more or new jobs and a better quality of the overall living environment (see also EEA, 2010). These co-benefits may also strengthen the long-term rationale of climate policy. However, decarbonization by 2050 asks for a long-term, strategic perspective that reveals long-term opportunities and requirements that need to be addressed today.

A long-term perspective draws attention to particular internal policy measures that need to be promoted. Fostering unproven, but necessary technologies in industry (Chapter 5) or investing in grid interconnections (Chapter 4) or new transport infrastructure and technology (Chapter 6) are necessary for responding to the challenge of climate change. Furthermore, long-term planning brings investment cycles in each sector in focus. In the power and industrial sectors, many investment decisions are taken on a much longer cycle (up to 30 years or more) than electoral cycles (typically 4–5 years) (see Chapters 2–5). In the buildings sector, low-energy renovations of the existing building stock require sustained efforts over decades (Chapter 7). Even in the transport sector, the time required for the development of new vehicle designs and production lines means changes towards low-carbon development require early and sustained policy action (Chapter 6). In turn, failing to implement policies that guide and enable investments with a long-term decarbonization perspective may risk havoc for the relevant sectors as the need for climate action grows and becomes more urgent.

Under these circumstances, a stable policy framework is required – that goes beyond targeted short-term policy intervention – in order to provide certainty to actors that long-term investments into decarbonization will pay off (Chapter 2). If policy decisions are to be valid into the long-term, we may need cross-party agreement on the importance of moving to decarbonization or even constitutional, or other legal, safeguards to ensure that future politicians and policymakers do not retract policies that lead to decarbonization (Dupont, forthcoming). If the long-term perspective of achieving decarbonization by 2050 is not central to policy planning, opportunities of low-carbon investments are missed, the risk of stranded assets increases, and valid long-term policy guidance does not emerge.

For developing external relations under decarbonization, long-term strategic planning entails opportunities but also brings out particular challenges. It opens the prospect of moving away from conflictual relations of vulnerability and security inherent in the geopolitical framing of energy relations (see Chapter 8) towards an emphasis on cooperation on low-carbon technologies and other fundamental 'European values' such as human rights and the rule of law (see Chapter 9). With strong, internal long-term planning it may prove easier to prioritize decarbonization objectives in external relations (with the freedom from dependence on fossil energy). However, it may prove challenging to shift relationships with Russia, the Caspian region, Norway and others away from reliance on fossil fuel imports (Chapters 8–11). The EU could actively prepare and pursue the shift in interactions with external energy partners that is implied by decarbonization, and also develop new partnerships, if the long-term 2050 perspective were taken better into account (Chapters 8–11; see also below).

By ensuring decarbonization is central to its external bilateral and multilateral relations, beyond climate negotiations only, the EU can try to influence other jurisdictions to take action and ensure coherence across relations. It is in the EU's interest to push for international action, both through involvement and leadership in international negotiations and through its bilateral external relations, to address fears about the EU going it alone and the implications this may have on the competitiveness of the EU's industry. The EU can use external relations to export the decarbonization agenda for its own benefit and that of its partners. Such a strategy requires long-term planning in advance of all bilateral and multilateral engagements, with EU efforts to put decarbonization at the centre of any agenda.

4. Sectors require tailor-made approaches

Taking the sectoral analyses in this volume together, it also emerges that there is no single one-size-fits-all solution. Tailor-made sectoral approaches and policy frameworks are required. Different sectors of the economy require specific policy responses, in particular given the varying stages of development of technological solutions and the extent to which sectors can relocate operations in response to climate policy. However, it is important to note that sectors are interlinked. If the power sector fails to decarbonize on time, this will have significant implications on the ability of high energy-consuming sectors to decarbonize also. In turn, developments with respect to vehicles, grid infrastructure and buildings and energy efficiency will affect the power sector's ability to decarbonize.

As regards technological development, core carbon-free, renewable technologies are at an advanced stage of development in the power sector, although technological challenges exist with respect to handling an increasing amount of intermittent renewable electricity generation (grids, storage) (Chapters 2–4). In the buildings sector, many technologies required for decarbonization also exist (Chapter 7). In the transport sector, low-carbon technology is under development but needs to advance significantly (Chapter 6). In the energy-intensive industry, many technologies required for decarbonization still need to move from the conceptual stage to development and demonstration (Chapter 5). Technologies across the sectors may also face particular challenges of societal acceptance. Technological improvements in industrial processes, for example, may not necessarily interfere with the everyday life of a community, whereas building large wind turbines can disturb (especially rural) communities unused to such installations.

With respect to international flexibility, industry may present the main challenge since it produces internationally traded goods and can thus, in principle, relocate production and production capacity over time (which limits the possibilities for European policymaking to impose regulation/emission limits). In contrast, power generation, transport and buildings are located in Europe and cannot easily evade European climate policy, even though some of the businesses involved (especially car manufacturers) can point to effects on their international competitiveness.

Against this backdrop, policy measures and frameworks required in each sector to advance towards decarbonization face particular conditions and have to address varying challenges. Policy measures for industry are especially required to encourage research into innovative breakthrough technology and its subsequent demonstration and deployment. Given the relative international flexibility of production and the costs involved in developing technologies, targeted incentive schemes may be required (Chapter 5). Policy measures in the building sector, in contrast, primarily need to become more ambitious and stringent. Much of the technology to improve the performance of buildings is available, but the policy framework is an insufficient driver of change. Given member state reservations, incentives schemes may again be required to complement stronger building standards and stricter renovation roadmaps (see Chapter 7). In the power sector, steps towards the greater expansion of renewable energy and a long-term downward trajectory towards decarbonization are required together with research on and investment in infrastructure and storage that may be driven both by standards and incentives (Chapter 2–4). In the transport sector, finally, policies need to nurture different technologies and approaches, including vehicle emission standards, measures to enable and incentivize modal shift, infrastructure development and low-carbon public transport and integrated city planning (Chapter 6; Ten Brink, 2010). Interestingly, the demand for research is affected by progress towards decarbonization in the power sector: with lack of progress here, alternative low-carbon technological options to electrification in transport will then be required.

In spite of the variance highlighted, some commonalities across the sectors also emerge from the analysis. Further research into the technological solutions and their implications on society would appear to be required in all sectors – albeit to varying degrees. Research is required for innovative breakthrough technologies in industry, vehicles and infrastructure in transport, electricity grids and energy storage technology in the power sector and improvements in existing technology in the buildings sector. This may best be accompanied by research on the societal impact of such new technology, including on the economic, social and political implications of the infrastructural changes required by decarbonization (see also below). It is also apparent that targeted incentives may play an increasingly prominent role for reasons that vary between the sectors (such as global reach of industry, overcoming incentive-barriers in buildings and power infrastructure development).

As with sectoral policy frameworks, differentiation and a country-specific approach are also required in external relations. The challenges and opportunities presented by decarbonization will not necessarily be the same for all external partners. Where relations with Russia (with some political creativity) can pivot to promote trade in renewable electricity and biofuels, there is little potential for such trade with the Caspian region at present. Broader political relations with these countries and regions can raise challenges and opportunities for focusing on issues that may be either more or less antagonistic. Relations with Norway are far less politically sensitive, but decarbonization will nevertheless present challenges for the EU-Norway partnership more broadly, as Norway continues its reliance on fossil fuels as its major export (see Chapters 8–11; see below).

In sum, it is clear that there is no single solution to the development of internal policy sectors and of external relations on the road to decarbonization. The EU must move forward on a sector and case-based approach, while taking into account that action in one area is likely to have knock-on effects on actions required in other areas.

5. The ETS is not a silver bullet

From the above discussion it emerges that the EU ETS is in itself insufficient as the central climate policy instrument of the EU. It has been the flagship climate policy of the EU since the first Directive was adopted in 2003 (2003/87/EC). However, the sectoral analyses in this volume indicate that the EU ETS needs to be complemented and embedded in a broader policy framework in order to advance sufficiently towards decarbonization. The ETS provides neither sufficient incentive to the sectors it already covers to invest in more long-term solutions to reduce emissions, nor does it have the potential to make the decisive difference in this respect for the sectors not (yet) covered by it.

In the industry and power sectors already covered by the ETS, the ETS has so far been insufficient to drive decarbonization, particularly because of relatively low carbon prices. The carbon price fluctuation and drop is due to several factors, both internal and external to the design of the system. The financial and economic crises since 2008 played a significant role in the falling prices, due to lower demand for emission permits (Carbon Market Watch, 2014; Wyns, 2015). But the sub-optimal functioning is also due to flaws in the design of the ETS, and especially, an over-allocation of free emission permits, a high inflow of international credits and the inability to adjust supply to decreasing demand (Wettestad, 2011; Wyns, 2015). Later revisions of the ETS reduced the free allocation of permits, but many industries still receive free allocation due to fears of 'carbon leakage' or industry relocation outside the EU (see Chapter 5). In combination, free allocation of permits and falling demand has left little incentive for investment in efficiency or low-carbon energy measures to mitigate climate change in both the power and the industry sectors over the longer term. While GHG emissions have fallen in the EU, long-term structural changes for decarbonization that a well-functioning ETS could have promoted have not yet materialized.

At the same time, the analysis in Chapters 3–5 also suggests that even a well-functioning ETS would not be able to induce all the changes and investments required. Improving the functioning of the ETS is under discussion, including the proposal for a market stability reserve (to balance oversupply in the market) (Carbon Market Watch, 2014). While a higher price for emission permits may make a significant difference in power production itself, it is difficult to see how high prices would, without further political guidance and under conditions of unbundling of networks from power production (Chapter 2), translate into sufficient incentives for adequate grid and infrastructure development. Similarly, any permit price under the ETS is likely to be insufficient to provide incentives for investing in unproven breakthrough technologies in energy intensive industries because of very long lead times and high costs (also when compared to moving production elsewhere). Even with a functioning ETS, decarbonization in the sectors covered by it (power production and industry) requires employing a broader set of policy instruments.

The same may hold even more clearly for the sectors analysed in this volume that are not (yet) included in the EU ETS. In both the transport and building sectors, costs of emissions play a subordinate role and are overshadowed by other barriers that will not be overcome by signals from pricing GHG emissions. Pricing GHG emissions in transport by including the sector in the ETS, would add only a few cents to the cost of fuel per litre, providing little incentive for car manufacturers to put low- or zero-emissions vehicles on the market (and for consumers to buy them). Similarly, putting a price on carbon will not resolve lack of appropriately trained personnel or lack of incentives for low-carbon renovations emanating from the split interests between owners and tenants in the building sector. Integrating these sectors into the ETS would most likely generate little more than an incremental incentive for short-term savings in energy consumption to reduce emissions, but is unlikely to drive long-term coordinated policy development and investment in new technologies and societal responses.

In general, it seems that the requirement to decarbonize the EU beyond the power sector calls for a mix of policy tools and measures that respond to the long-term research, planning, investment and infrastructural needs, rather than promoting incremental short-term improvements. While the ETS may continue as one central plank of EU climate policy, other policy measures need to be employed to address barriers that are not price-related or where the price signal itself would be too weak to induce the change required.

6. Sustained political commitment is necessary

Across all the chapters in this volume, deep political commitment to the decarbonization goal was highlighted as playing a significant role in whether or not decarbonization is likely to be achieved. This is true for both internal sectoral policy development and the evolution of external relations. Unless political commitment to decarbonization is consistent and at a high level, other (short-term) concerns are likely to take priority. A litmus test for such political commitment is that it is translated into meaningful long-term planning and strategizing towards the 2050 objective. Overall, the importance of political commitment is hardly surprising as it reconfirms earlier findings of the literature on environmental and climate policy integration (Jordan & Lenschow, 2010; Chapter 1).

In the development of internal EU policies, fluctuating political commitment has resulted in instability in the regulatory framework and uncertainty for investors. The level of ambition towards decarbonization in sectoral policies was tempered in the late 2000s by the economic and financial crises and the ensuing crisis in the EU more broadly. For the energy sector, this lowering of political commitment resulted in delayed action (e.g., Chapter 2). As argued in Chapter 1 and above, climate policy development in the EU has since lost dynamism (with, for example, ongoing problems with the EU ETS). The dwindling of sustained high political commitment to decarbonization has arguably delayed progress in closing the gap towards long-term requirements for decarbonization in most of the sectors investigated. Especially where sectors have yet to initiate the major structural changes towards a long-term trajectory to decarbonization (such as in industry, see Chapter 5), the tempered political commitment implies delays that are hard to catch up on. Without high political commitment to decarbonization within sectors, some of the sector-specific challenges may not be overcome.

Awareness and recognition of functional interactions of a policy area or external strategy with the decarbonization goal seem to be an important precondition for sufficient political commitment to sectoral change. As long as the potential for synergy or conflict between a policy sector or an external strategy and decarbonization are neglected, it is difficult to imagine that political commitment for such change will emerge (even if general political commitment to combating climate change is high). The contributions to the volume provide evidence that the awareness of functional interactions has grown and broadened over time. For example, the importance of electricity grids (Chapter 4), active long-term industrial policy (Chapter 5) or transport policy beyond fuel standards for cars (Chapter 6) for decarbonization by 2050 has only been recognized and understood over time. In the case of external energy relations, the interlinkages are only beginning to be acknowledged, if at all (Chapters 8–11). The broad recognition of functional interactions and its translation into political commitment may not be facilitated by the cross-cutting nature of climate change, as decarbonization requires knowledge and political action on a wide range of sectors and external relations. The chapters in this volume indicate that awareness of interlinkages has been expanding, but remains incomplete. This may imply a joint task of both researchers and policymakers to deepen further the understanding of interlinkages.

A further boon to levels of political commitment to decarbonization can come from the engagement of environmental and climate stakeholders that push policymakers. Where important policy initiatives have been successfully adopted – including the EU ETS, the Renewable Energy Directive, the CO_2 and Cars Regulation and others (see Table 1.1) – the engagement of environmental and climate stakeholders has usually been high. In contrast, where policy development has remained lagging – including in buildings policy (see Chapter 7), transport policy more broadly (Chapter 6), industrial policy (Chapter 5) and external relations (Chapters 8–11) – the involvement of such stakeholders has usually also been lower. Two particular barriers can be identified for enhanced engagement: resource constraints on the side of relevant societal stakeholders and limited possibilities for input in the policymaking process. Potential for improvement thus exists on the side of stakeholders themselves to take better account of the functional overlaps, allocate resources and advocate long-term political commitment to decarbonization. Addressing structural constraints to accessing policymaking, for example in EU external relations, may also require attention.

7. Competence division shapes the governance of decarbonization

The division of competences between the EU and its member states, both regarding policymaking and implementation, shapes EU climate and energy governance towards decarbonization. In some areas of EU climate and energy policies, including energy efficiency standards for products and CO_2 standards for cars, competence clearly rests at the EU level. In many other areas, however, competences are mixed or shared between the EU and national levels so that policymaking at both levels is possible and sometimes required. While this increases complexity, the implications entail both advantages and complications towards decarbonization.

Mixed or shared competences allow room for initiatives by member states (and within member states) that may circumvent and overcome lack of progress at the EU level. Member state initiatives to develop and implement policies may facilitate experimentation and create front runners on certain solutions. For example, progress on renewable energy in the EU has very much relied on individual member states designing and introducing support schemes. These member state initiatives were subsequently important drivers of the development of policy in support of renewable energy at the EU level, especially the 2009 Renewable Energy Directive. Decentralized member state competence can thus create dynamics that drive EU-level policies forward – in a similar way as they can drive the EU's international climate policy (Schreurs & Tiberghien, 2007).

However, progress in EU climate and energy governance at times requires EU level action or cross-border coordination to exploit the potential for joint action, to address and overcome barriers or to make laggards move. In these cases, shared competences can hinder timely and sufficient progress. For example, cross-border infrastructure developments, especially in the power and grid sectors, require coordination among individual member states, but also including regional and local actors (see Chapters 2–4). In the industry and transport sectors, some sort of shared vision and joint action of EU member states could help overcome bottlenecks in policy development and implementation (Chapters 5 and 6). For the buildings sector, the division of competences has played a crucial role in watering down EU policy to improve the performance of buildings as many member states have argued for buildings policy being made at lower levels of governance. Here, member state action has not (yet) created reinforcing upward dynamics, which has resulted in unequal and insufficient progress across member states (Chapter 7). Where the guarding of competences is impeding progress in EU level regulation, the creation of targeted incentive schemes may provide some way forward (see Chapters 4–7).

In external relations with energy partners, shared competence between the EU and its member states does not necessarily facilitate streamlining the decarbonization objective into policy. While foreign policy falls into the hands of both the member states and the EU (and arguably even more in those of the member states), effective coordination is needed to ensure a united external message on decarbonization and climate change. In the cases investigated, there is little evidence of a systematic and sustained streamlining of the decarbonization objective into relations with external energy partners and, consequently, of coordination within the EU to this effect (see Chapters 8–11; see also below). Broader political contexts have drowned out the EU's decarbonization message, likely reinforced by a lack of coordination among the EU and its member states.

Overall, the division of competences within the EU, and more generally the multi-level setup of EU governance, constitute an important area for further research. While this volume has primarily focused on the EU level, the dynamics of multi-level interactions with EU climate and energy governance deserve further investigation (see below).

8. Low integration of decarbonization into external relations

The internal EU objective to decarbonize by 2050 is not a centrepiece of the EU's external relations with energy partners (Chapters 8–11). As a clear long-term goal of the EU, with several short- and medium-term implications it seems logical that EU external energy relations would reflect this internal objective to decarbonize. In the cases examined in this book, long-term strategizing around the decarbonization goal is not evident. In the case of Russia, some common exploration of a decarbonized future has taken place, but seems to have primarily remained at the conceptual level without significantly affecting material relations that remain dominated by fossil fuel trade and broader issues (Chapter 10). With respect to Norway, some modest initiatives to develop relations towards expanded use of renewable energy are underway, but the major basis of bilateral relations remains on fossil fuels (Chapter 11). In the case of the Caspian region, decarbonization hardly seems to have been factored into relations at all (Chapter 9).

On the one hand, putting decarbonization at the core of external relations presents the EU with certain strategic challenges. It implies a fundamental change of external relations and the interdependence relationship with existing energy partners away from a focus on fossil fuels. Diminishing demand for fossil fuels in the EU does not necessarily mean independence from current partners, but it does mean a significant rebalancing of the complex interdependence equilibrium that would rather be based on factors that shape relations with other non-energy partners (including international trade, security and others) (on complex interdependence, see Keohane & Nye, 1977). Such a shift is likely to require active management of potential tensions, especially where relations are precarious for other reasons (such as Russia, see Chapter 10). Where trade in fossil fuels has also provided the EU with influence as a major market, EU external relations and diplomacy will have to adapt to its changing patterns of influence.

On the other hand, and in managing the aforementioned challenges, decarbonization could represent an opportunity for external energy relations to move beyond the realm of geopolitical control over (fossil) resources (Chapters 8–11). Long-standing EU external relations based on fossil fuel trade limit the EU's room for manoeuvre when political relations become strained in other domains (for example, in the case of EU–Russia relations). Opportunities exist to move beyond these traditional relations. If decarbonization were at the heart of external relations, the EU could break free of challenging political developments with some of these third countries. In its relations with Russia, for example, the EU would be less encumbered with dependence on Russia for significant shares of fossil fuel supplies (see Chapter 10). In the Caspian region, the EU could shift relations away from seeking access to fossil resources towards being a pusher of democratic and human rights values (see Chapter 9). Furthermore, decarbonization could open opportunities for closer ties with other neighbouring countries, with respect to trade in renewable electricity, for example. This may temper the influence of major global powers in the EU's neighbourhood and tie economies in transition to a development path linked with the EU.

Overall, there has been little translation of the decarbonization goal into long-term strategies for developing external relations – probably because of the short-term, reactive nature of some of the EU's external relations and the lack of long-term planning as mentioned above. Relations with energy partners may be locked into a certain framing of energy resources as a source of (geopolitical) power (see Chapter 8). Bringing decarbonization into the centre of such relations has the potential to move the EU beyond this geopolitical understanding of energy relations and to highlight other EU priorities. Addressing and managing the resulting challenges and opportunities, and their country-specific forms, timely, consciously and in a strategic manner seems necessary to help smooth the transition and prevent any potential tensions from spiralling.

9. A rich research agenda

Based on the insights gained in this volume, we can identify a number of promising avenues for further research. This volume could only provide some answers to a few questions that the decarbonization perspective raises for EU internal and external policy and strategy. There is scope and need for much more research into the political, social, economic and technological changes that the transition to decarbonization requires and implies. Here we discuss five potential areas for future research on decarbonization and the EU, as we see them emerge from the analysis of this volume.

First, there is a dire need for further exploration of the challenges and opportunities that decarbonization in the EU (and beyond) entails for EU external relations. Chapters 8–11 of this volume suggest that rather limited analysis exists of the implications of the move to decarbonization on EU external relations with energy partners. It is likely that this finding holds also for other external energy partners not addressed in depth in this volume (including oil exporting countries, Northern Africa and others). Including partners that do not export significant fossil fuels to Europe extends the field of interest to be explored much further. With the technological and economic changes implied by decarbonization, relations with all international partners (including the United States, China, Japan, Brazil, South Africa, India and others) are likely to be affected and thus deserve exploration. More detailed and more encompassing knowledge about the prospect of external relations under decarbonization may provide a useful basis for investigating the scope for proactive political management of the opportunities and challenges arising. Building on this, further exploration of the role of the EU in a decarbonized world order, from a more encompassing long-term perspective, could be carried out.

Second, future research on decarbonization in EU multi-level governance holds particular promise. While the discussions in this book focused on the EU level, the analysis showed the importance of the role of member states and the division of competences. Other levels of governance also play significant roles, including the sub-national and international levels (see Bulkeley & Newell, 2010). The interlinkages and interactions between the different levels deserve particular attention in future research. Under what circumstances and conditions should action best be taken at which level or levels? What drives the dynamics of multi-level interactions in EU climate and energy governance? How and under what circumstances does action at different levels reinforce or block each other? These are some of the questions that may provide starting points for a more in-depth understanding of EU multi-level governance towards decarbonization and may also deliver policy-relevant outcomes as to where to take action.

Third, and relatedly, research should explore what policy mixes and tools work best in which sectors, and why, from a long-term decarbonization perspective. Such a sector-by-sector analysis should be embedded and informed by an investigation of how different sectoral developments interact, and need to interact, under decarbonization. For example, while we need to know more about the best policy mixes (and governance levels) to employ in the industry, power and transport sectors, we also need to understand how progress in the power sector, and the form it takes, will shape the prospect for decarbonization in the industry and transport sectors, and vice versa. In other words, sectoral analyses eventually need to be integrated. One step further in this integration is the exploration of how decarbonization can align and interact with other prominent policy objectives, including economic development, industrial innovation or energy security.

Fourth, research on the impacts of decarbonization on the vulnerable in society – not discussed in this volume – seems a logical extension to inform long-term planning on how to cope with the transition. This may link to the economic costs and benefits of moving towards decarbonization, but research that assesses the many knock-on benefits of decarbonization (improved health, clean air, new jobs, improved quality of the environment) should go hand in hand with assessments of monetary costs of a transition. The distribution of these costs and benefits then comes into focus in the further exploration of the social dimension of decarbonization, which seems fundamental for maintaining and reinforcing societal support for decarbonization and legitimacy for policy decisions to promote the transition.

Fifth, conceptual considerations of the interrelationship between a long-term decarbonization perspective and the quality of democracy in the EU would prove fruitful. Long-term planning for decarbonization may have both positive and negative effects on democratic systems in the EU that focus on short-term electoral cycles and changing policy preferences. Cross-party agreement or constitutional provisions that ensure decarbonization is achieved by 2050 may affect future electorates' abilities to influence the policy direction of their representatives. Conversely, failure of today's elected officials to implement sufficiently ambitious policies to achieve decarbonization may also negatively impact the range of policy options available to future electorates and future policymakers. Such a failure could negatively affect future generations' ability to sustain a reasonable quality of life, as the impacts of unabated climate change unfold. Research on reconciling long-term policy planning with short-term democratic electoral cycles is thus required to enhance understanding on the benefits and trade-offs for democratic representation and legitimacy of moving towards decarbonization.

In sum: Opportunities and challenges ahead

To summarize the main lessons already discussed, we can highlight that decarbonization represents an opportunity and a challenge for both the internal policy development and external strategic evolution of the EU. Recognizing both the challenges and opportunities can help the EU prepare and manage the required shift in internal sectoral policies and external relations in a timely fashion. With the recognition of the opportunities and challenges, policymakers can work towards framing decarbonization in terms of long-term gains and overlapping benefits, such as improving quality of life more generally and promoting other 'European' values.

The EU has made important progress both in terms of reducing GHG emissions and developing a comprehensive policy framework. At the same time, it needs to accelerate its GHG emission reductions and step up its climate and energy policies. Priorities include, next to a structural reform of the EU ETS, the further development of sectoral targeted policy mixes (with an increased emphasis on incentives and support) and the full integration of the decarbonization objective into all relevant policy domains. In developing the climate policy framework, due attention needs to be paid to multi-level linkages and how the different governance levels can synergize and reinforce each other. Moving towards decarbonization promises to make European society, economics and politics fit for the future, including by eliciting innovation, moving on the energy transition and avoiding stranded investments and carbon lock-in. It needs to be supported by knowledge about and awareness of functional interlinkages between the policy domains, strong societal support and sustained political commitment.

As regards external policies, the EU still has much to do to analyse the implications of its decarbonization agenda for the development of external relations and to adapt its external policy strategies accordingly. Doing so promises to highlight new opportunities, for example, building or renewing partnerships based on the use of renewable energy. It should also enable the EU to move beyond the geopolitical imprint of current fossil fuel relations on important international partnerships and to recalibrate international interdependence based on trade and other considerations. It might also make space for projecting internationally the Europe's contribution to fighting climate change and for demonstrating that the decarbonization transition can be managed while advancing social and economic development.

Future research thus meets a fascinating field of change. More research on both the internal and external dimensions is required and promises high returns. Advancing the related research agenda can significantly improve our understanding of multi-level governance in the EU, but also possesses a high potential to inform the management of the decarbonization transition.

References

BPIE (2011) *Europe's Buildings under the Microscope: A Country-by-Country Review of the Energy Performance of Buildings* (Brussels: Buildings Performance Institute Europe).

Bulkeley, H. & Newell, P. (2010) *Governing Climate Change* (London: Routledge).

Carbon Market Watch (2014), *What's Needed to Fix the EU's Carbon Market? Recommendations for the Market Stability Reserve and Future ETS Reforms*, www.carbonmarketwatch.org, date accessed 9 December 2014.

Dupont, C. (Forthcoming) *Climate Policy Integration into EU Energy Policy*, (London: Routledge).

Eckersley, R. (2012) 'Moving Forward in the Climate Negotiations: Multilateralism or Minilateralism?', *Global Environmental Politics*, 12(1), 24–42.

EEA (2010) *The European Environment – State and Outlook 2010: Synthesis* (Copenhagen: European Environment Agency).

EEA (2014) *Trends and Projections in Europe 2014: Tracking Progress towards Europe's Climate and Energy Targets for 2020* (Copenhagen: European Environment Agency).

European Commission (1998) 'Energy Efficiency in the European Community – Towards a Strategy for the Rational Use of Energy', COM(1998) 246.

European Commission (2011a) 'Communication from the Commission: Energy Roadmap 2050', COM(2011) 885/2.

European Commission (2011b) 'Communication from the Commission: A Roadmap for Moving to a Competitive Low Carbon Economy in 2050', COM(2011) 112.

European Commission (2011c) 'White Paper – Roadmap to a Single European Transport Area. Towards a Competitive and Resource Efficient Transport System', COM(2011) 144.

European Commission (2014) *EU Transport in Figures – Statistical Pocketbook 2014* (Luxembourg: Publications Office of the European Union).

European Council (2009) 'Presidency Conclusions', Document 15265/1/09, 29 and 30 October.

European Council (2014) 'Conclusions', Document EUCO 169/14, 23 and 24 October.

Hill, N., Brannigan, C., Smokers, R., Schroten, A., van Essen, H. & Skinner, I. (2012) 'Developing a Better Understanding of the Secondary Impacts and Key Sensitivities for the Decarbonisation of the EU's Transport Sector by 2050. Final Project Report', http://www.eutransportghg2050.eu/cms/assets/Uploads/Reports/EU-Transport-GHG-2050-II-Final-Report-29Jul12.pdf, date accessed 12 November 2014.

IPCC (2013) 'Summary for Policymakers', in T. F. Stoker, D. Qin, G.-K. Plattner, M. Tignor, S. K. Allen, J. Boschung, A. Nauels, Y. Xia, V. Bex & P. M. Midgley (eds.), *Climate Change 2013: The Physical Science Basis. Contribution of Working Group I to the Fifth Assessment Report of the Intergovernmental Panel on Climate Change* (Cambridge: Cambridge University Press).

Jordan, A. & Lenschow, A. (2010) 'Environmental Policy Integration: A State of the Art Review', *Environmental Policy and Governance, 20*(3), 147–158.

Keohane, R. O. & Nye, J. (1977) *Power and Independence: World Politics in Transition* (Boston, MA: Little, Brown).

Neuhoff, K., Acworth, W., Dechezleprêtre, A., Dröge, S., Sartor, O., Sato, M., Schleicher, S. & Schopp, A. (2014) *Staying with the Leaders: Europe's Path to a Successful Low-Carbon Economy* (London: Climate Strategies).

Oberthür, S. & Wyns, T. (2014) 'Paris Climate Agreement 2015: EU Needs to Ensure "Signal" and "Direction"'. *IES Policy Brief, 2014*(8).

Schreurs, M. A. & Tiberghien, Y. (2007) 'Multi-level Reinforcement: Explaining European Union Leadership in Climate Change Mitigation', *Global Environmental Politics, 7*(4), 19–46.

Ten Brink, P. (2010) 'Mitigating CO_2 Emissions from Cars in the EU (Regulation (EC) No 443/2009)', in S. Oberthür & M. Pallemaerts (eds.), *The New Climate Policies of the European Union: Internal Legislation and Climate Diplomacy* (Brussels: VUB Press), pp. 179–209.

Wettestad, J. (2011) 'EU Emissions Trading: Achievements and Challenges', in V. L. Birchfield & J. S. Duffield (eds.), *Towards a Common European Union Energy Policy: Problems, Progress, and Prospects* (New York: Palgrave Macmillan), pp. 87–111.

Wyns, T. (2015) 'Lessons from the EU's ETS for a New International Climate Agreement', *International Spectator*, doi:10.1080/03932729.2015.985941.

Index